Graphical representation

Graphs 1

A1

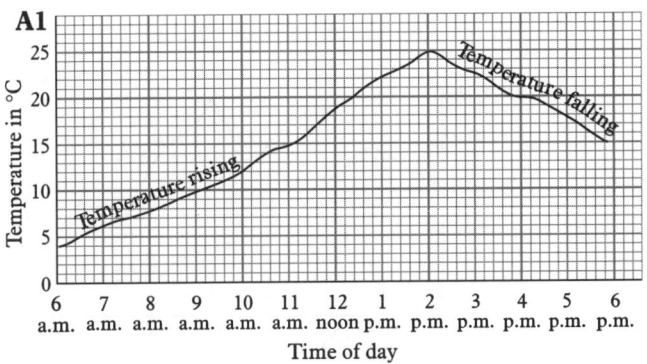

The temperature was highest at 2 p.m.
(It was about 25 °C.)

A2

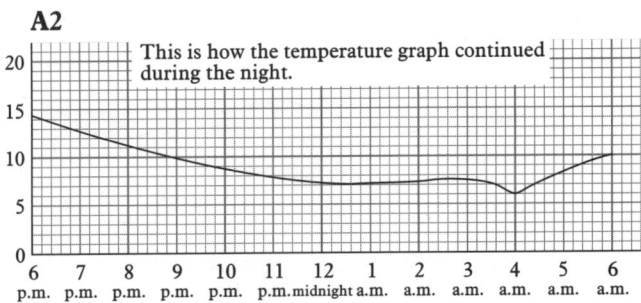

The temperature was lowest at 4 a.m.
(It was 6 °C.)

A3

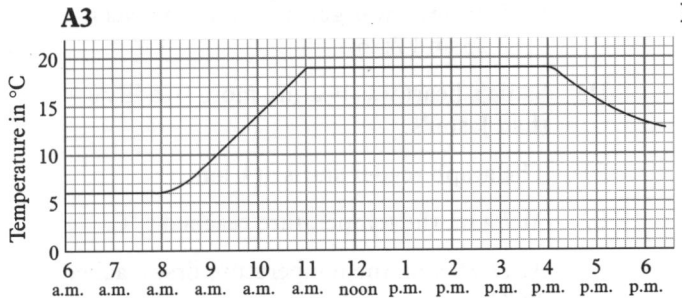

(a) The temperature began to rise about
8 a.m.
(b) The temperature stopped rising at
11 a.m. (11 o'clock in the morning).
(c) Between 11 a.m. and 4 p.m. it stayed
the same.

(d) The temperature started falling at
4 p.m. so it was about this time that the
heating was turned off.

A4 (a) This shows the
temperature rising
and then falling.

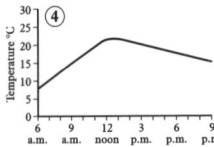

(b) This drawing shows
the temperature
rising at first and
then staying the
same.

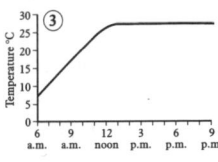

(c) This shows the
temperature staying
the same at first and
then falling.

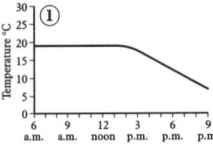

(d) The temperature
here stays the same
all the time.

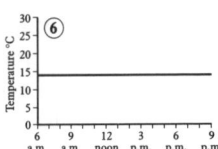

(e) This graph shows
the temperature
getting colder and
then warmer.

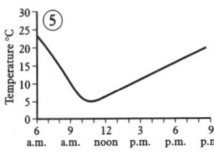

(f) The temperature is
getting warmer all
the time.

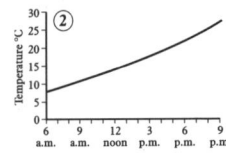

B1

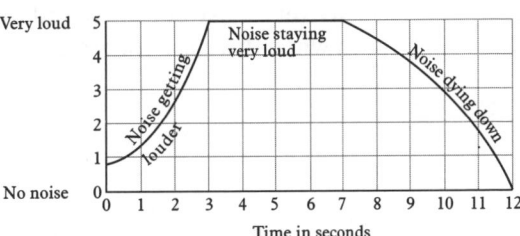

It stayed very loud for 4 seconds.

B2 It took 5 seconds for the noise to die down
▲ from very loud to nothing.

1

B3

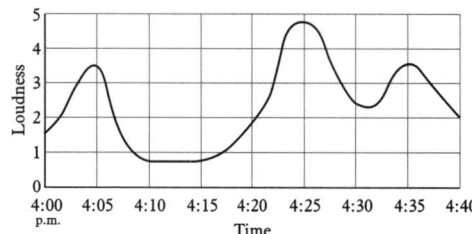

(a) The graph shows that 3 planes flew over Bill's garden.

(b) The second plane was the noisiest.

(c) The loudest plane was right over Bill's garden at 4:25 p.m.

(d) The quietest part was from 4:10 to 4:15.

B4 (a)

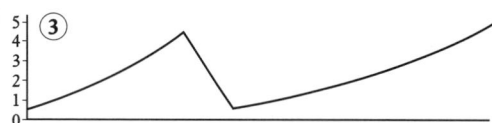

This graph shows the noise getting louder and then getting quieter as the teacher tells the class to be quiet. But the class then gets noisier again.

(b)

The graph shows the noise made by 3 loud explosions.

(c)

This shows the noise of a factory. It's usually very noisy but the sound dips down when there is a tea break.

C1

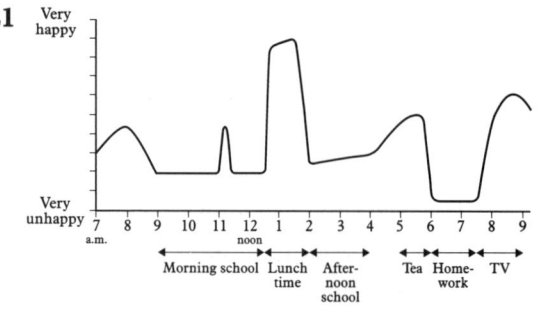

(a) Mary probably got less happy just before 9 a.m. because she was going to school.

(b) It was probably morning break between 11 a.m. and 11:15 a.m.

(c) Yes, she was happier at lunchtime than when she was watching TV.

(d) She seems to be most unhappy at homework time.

C2 Show your graph to your teacher.

D1

(a) Jason's heart was beating fastest at 4:19.

(b) His heart was beating slowest at 4:22.

D2

(a) Between 7:21 and 7:23 Sadia's heart was getting faster.

(b) Her heart was getting slower between 7:23 and 7:26.

D3 Sadia's heart was beating more steadily than Jason's.

E1 The graph reader starts at 6 a.m.

E2 At 7 a.m. the temperature reached 5 °C.

E3 At 11:30 a.m. the temperature first reached 20 °C.

E4 At 5 p.m. the temperature had gone down to 10 °C.

E5 At noon on Monday the temperature was **10 °C**. Between noon and 3 p.m. the temperature was **rising**. The highest temperature was **15 °C**. It occurred at **3 p.m**. After that the temperature was **falling** for **11** hours. The temperature was lowest at **2 a.m.** on **Tuesday**. It was **2 °C**. Then it started to **rise**.

E6 (a) On Monday at 8 p.m. the temperature was 7 °C.
(b) Between 8 p.m. and 9 p.m. the temperature fell.
(c) The temperature rose between 4 a.m. and 5 a.m. on Tuesday.

E7 (a) On Monday the temperature was above 10 °C for $6\frac{1}{2}$ hours.
(b) On Tuesday the temperature reached 8 °C at 8:30 a.m. (this is half past eight in the morning).

E8 The temperature rose quickly after 7 a.m. on Tuesday probably because this was the time that the sun rose.

E9 Temperatures like those on the graph can happen at any time of the year in Britain, but it is most likely to be spring or autumn.

F1
▲

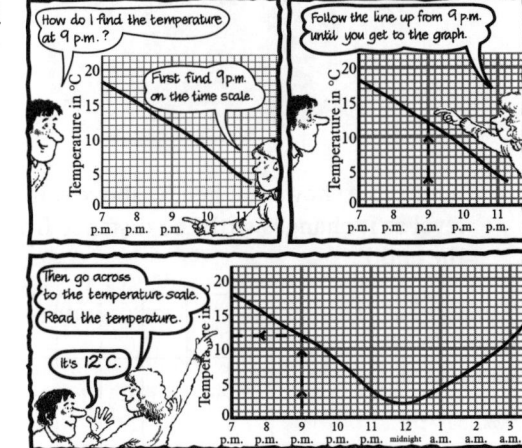

(a) At 8 p.m. the temperature was 15 °C.
(b) The temperature are 11 p.m. was 4 °C.
(c) At midnight the temperature was 2 °C.
(d) By 2 a.m. in the morning the temperature was 7 °C.
(e) At 3 o'clock in the morning the temperature was 11 °C.

F2

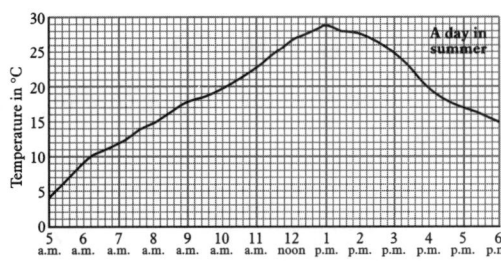

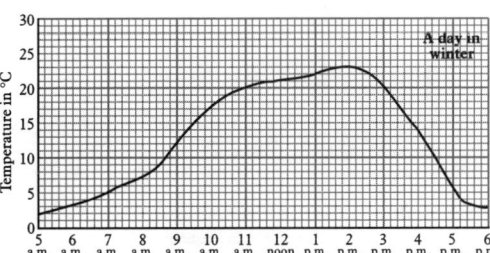

For the day in summer:
(a) at 5 a.m. the temperature was 4 °C
(b) at 6 o'clock in the morning the temperature was 9 °C
(c) between 5 o'clock and 6 o'clock in the morning the temperature went up by 5 °C
(d) the temperature was highest at 1 p.m.
(e) the highest temperature was 29 °C
(f) at 5 p.m. the temperature was 17 °C

F3 In winter:
(a) at 5 o'clock in the morning the temperature was 2 °C
(b) at 7 a.m. it was 5 °C
(c) 7 °C was the temperature at 8 a.m.
(d) between 7 and 8 o'clock in the morning the temperature had risen by 2 degrees
(e) between 9 a.m. and 10 a.m. the temperature rose by 5 degrees
(f) the highest temperature was 23 °C
(g) it occurred at 2 p.m.

F4 For the winter's day:
(a) at 11 a.m. the temperature was 20 °C
(b) at 3 p.m. the temperature was 20 °C
(c) at 4 p.m. the temperature was 14 °C
(d) between 10 o'clock in the morning and noon the temperature rose by 4 degrees
(e) between 3 p.m. and 5 p.m. the temperature fell by 13 degrees

Drawing curves

Show your graph of the hanging cord to your teacher.

A1

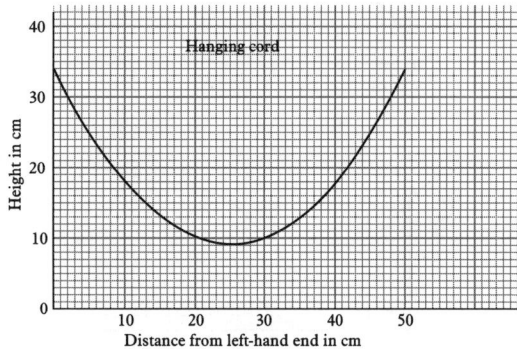

(a) The ends of the cord were 50 cm apart.
(b) The lowest point on the curve was 9 cm above the ground.
(c) The left-hand end was 34 cm high.
(d) 10 cm across from the left-hand end the cord was 18 cm high.
(e) There are two points which are 30 cm up from the ground. These are 2 cm across and
(f) 48 cm across.

B1

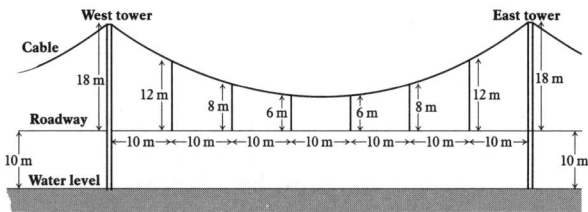

(a) The distance between the west and east towers is 70 m.
(b) The top of the west tower is 18 m above the road.
(c) The top of the east tower is 28 m above the water level.

(Remember, you need to add on 10 metres to find distances above the water level.)

B2 Here is a table showing the height of the cable above the water level.
(Don't forget that you need to add on 10 metres to find distances above the water level.)

Distance across from the west tower, in metres	0	10	20	30	40	50	60	70
Height of cable above water level, in metres	28	22	18	16	16	18	22	28

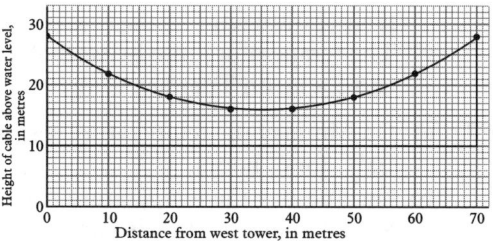

B3

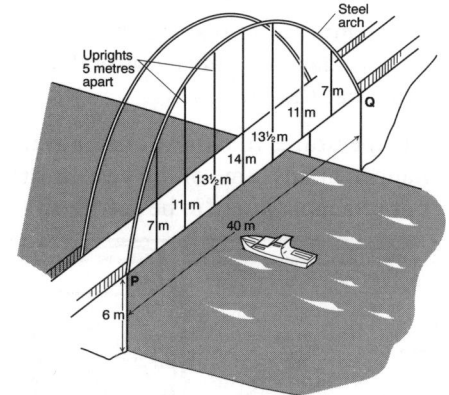

This table shows the height above water level as it changes with the distance from the end P.

Distance from end P, in metres	0	5	10	15	20	25	30	35	40
Height of arch above water level, in metres	6	13	17	$19\frac{1}{2}$	20	$19\frac{1}{2}$	17	13	6

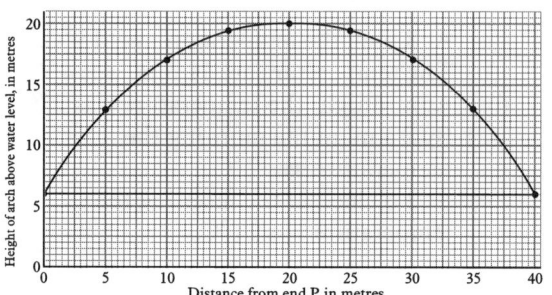

B4 A man who is 16 metres above the water level is between 8 and 9 metres across the bridge.

C1 Your curve should look like this.

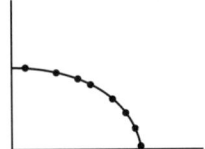

C2 Show your curve to your teacher.

C3 Show your curve to your teacher.

D1 The curve looks like this.

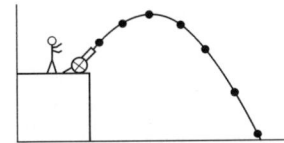

E1

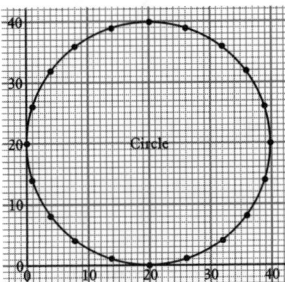

The curve is a circle.

E2

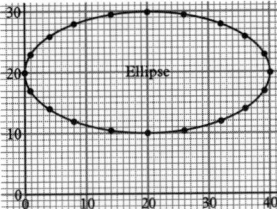

E3

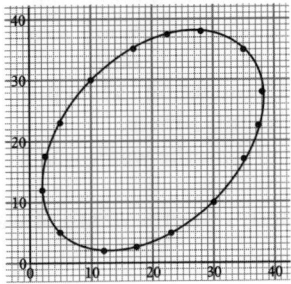

Graphs 2

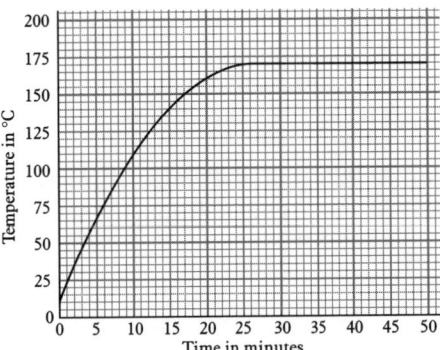

A1 (a) At the start the oven temperature was 10 °C.
△
  (b) The highest temperature reached was 170 °C.

  (c) It took 25 minutes to reach its highest temperature.

A2

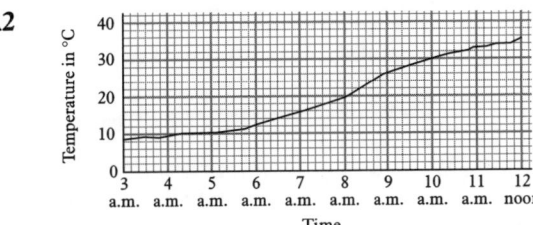

  (a) Each small square stands for 2 degrees.

  (b) Each small square on the time axis stands for 10 minutes.

  (c) At 8:40 the temperature was 24 °C (8:40 is 4 small squares past 8 a.m.)

  (d) At 9:30 a.m. the temperature was 28 °C.

  (e) 16 °C   (f) 10 a.m.

A3

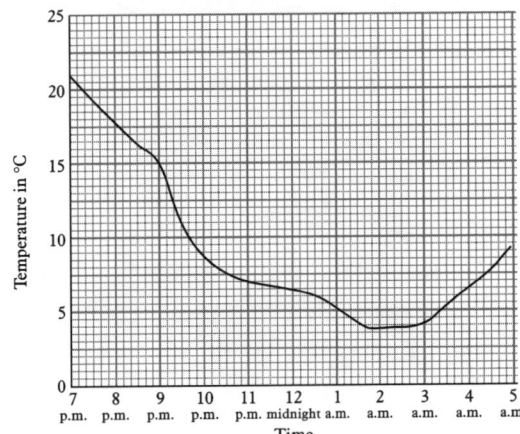

(a) Each small square on the temperature axis stands for $\frac{1}{2}$ degree (or 0·5 degree).

(b) At 9 p.m. the temperature was 15·5 °C (or $15\frac{1}{2}$ degrees).

(c) At 3:30 a.m. the temperature was 5 °C.

(d) The lowest temperature on the graph is 3·5 °C.

(e) At 9:20 p.m. the temperature was 13 °C.

(f) At 1 a.m. and 3:40 a.m. the temperature was 5·5 °C.

B1–B6

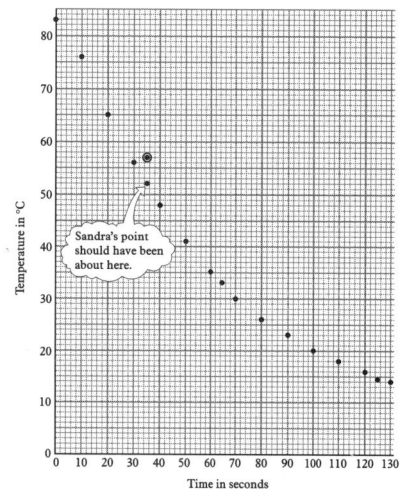

B2 A good guess for the temperature after 60 seconds is about 35 °C.

B4 After 30 seconds a good guess is about 55 °C.

B7 and B8 Show your graph to your teacher.

C1 and C2

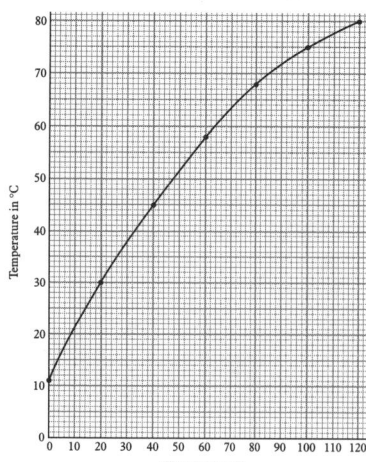

C2 (b) After 50 seconds the temperature would be about 52 °C.

C3

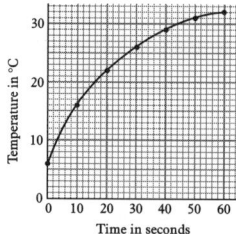

D1 This is the labelled graph.

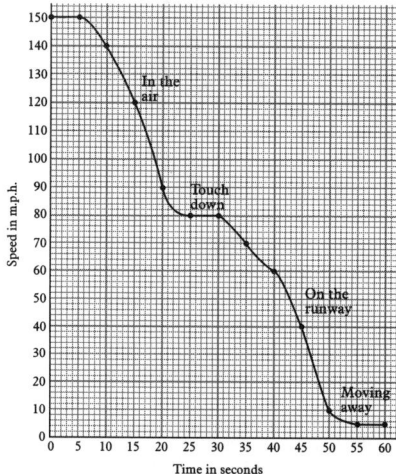

E1 (a), (c) Check your graph against this one.

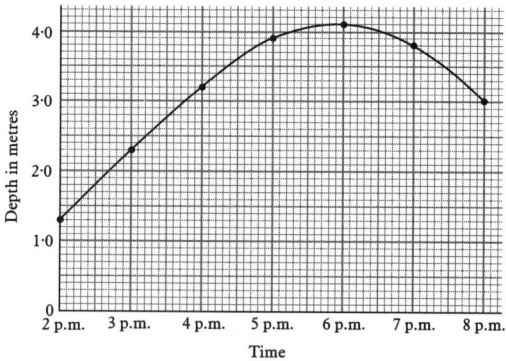

(b) 1 small square on the 'depth' axis stands for 0·1 metre.

(d) At 4:30 p.m. the water was 3·6 metres deep.

(e) 3·0 metres was the depth of water at 3:45 p.m.

F1 Sophie has not marked 4 weeks and 5 weeks on the time axis. So her time axis is not numbered in equal steps.

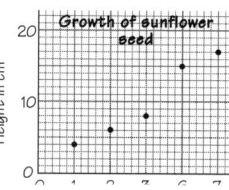

6

F2

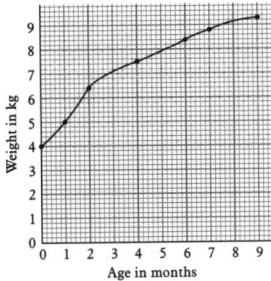

F3 Here is the graph showing how the candle's height changes with time.

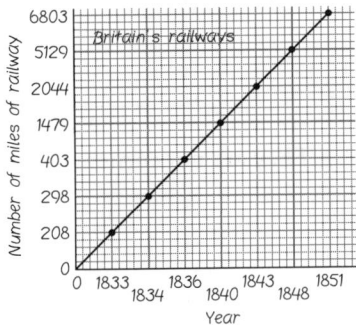

F4

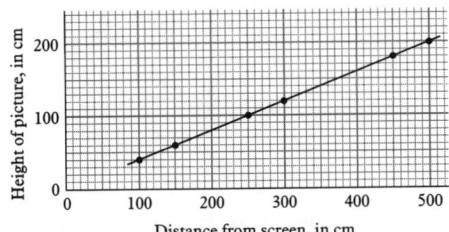

The axis showing the year is not marked in equal steps.
The axis showing the number of miles of railway is not marked in equal steps either.

G1

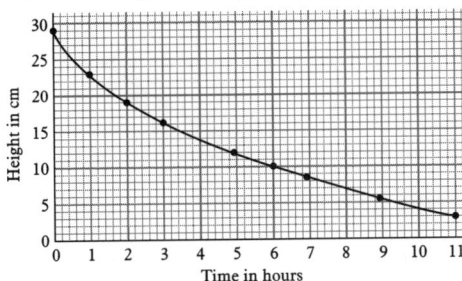

(a) The points are in a straight line.
(c) When the projector is 350 cm from the screen the picture is 140 cm tall.
(d) When the picture is 50 cm tall the screen is 125 cm away from the projector.

G2

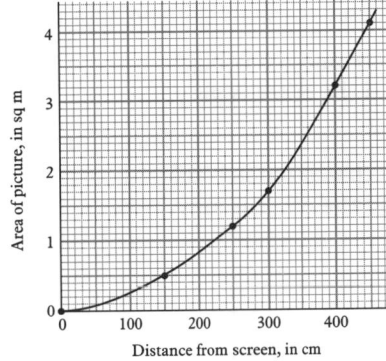

The projector will be 230 cm from the screen for a picture with area 1 sq m.

H Show your graph to your teacher.

Co-ordinates 1

A1

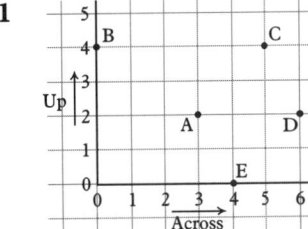

The co-ordinates of C are (5, 4), D (6, 2) and E (4, 0). (Remember, the across co-ordinate is first and the up co-ordinate second.)

A2 (a), (b)

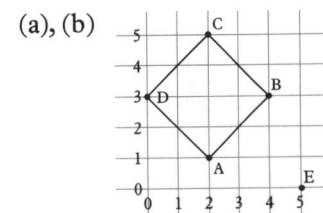

(b) Is your point D in the same position as this one?
(c) The co-ordinates of D are (0, 3).

A3 (a), (b), (d)

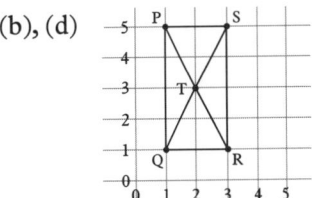

(c) The co-ordinates of S are (3, 5).
(e) The co-ordinates of T are (2, 3).

7

B1 There are 6 zeds in the diagram.

B2 Here one zed is black and the other is grey.

B3 Here one zed is black and the other is grey.

B4

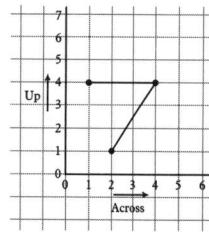

The fourth point could have either of these co-ordinates: (5, 1) or (3, 7).

B5

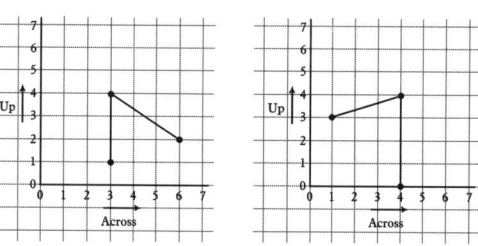

(a) The end of the zed could be either (6, 5) or (0, 3).

(b) The end of the zed could be either (7, 1) or (1, 7).

B6 See **B5**.

B7

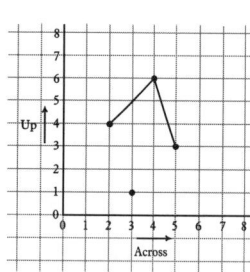

The end of the zed could be either (7, 5) or (1, 7).

C1

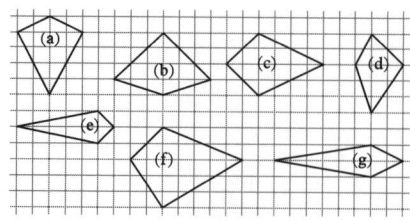

(a), (b), (c), (e) and (g) are kites. (Remember, a kite must have a line of reflection symmetry.)

C2 (a)

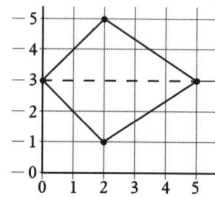

(b) This is a kite.

C3 (a)

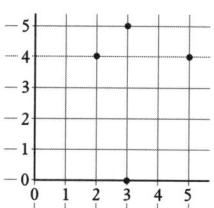

(b) They are not corners of a kite.

C4 (a), (b)

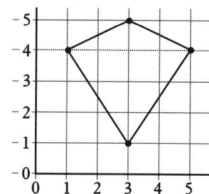

(c) The co-ordinates of the fourth corner are (5, 4).

C5 (a), (b)

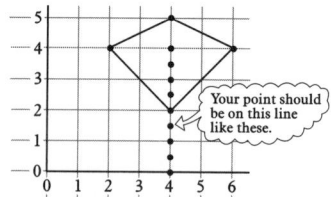

Your point should be on this line like these.

(c) The co-ordinates of the other corner are (4, 2).

(d) No, there are others.

(e) The other point could be: $(4, 0)$, $(4, \frac{1}{2})$, $(4, 1)$, $(4, 1\frac{1}{2})$, and so on.

D1 (a), (b)

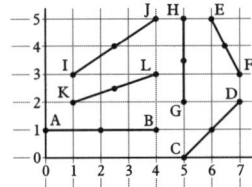

(c) The table shows the mid-points of the lines.

Line	Mid-point
AB	$(2, 1)$
CD	$(6, 1)$
EF	$(6\frac{1}{2}, 4)$
GH	$(5, 3\frac{1}{2})$
IJ	$(2\frac{1}{2}, 4)$
KL	$(2\frac{1}{2}, 2\frac{1}{2})$

D2 (a)–(d)

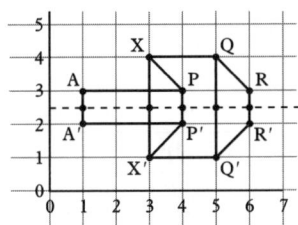

(e) Here are the co-ordinates of each of the mid-points:

$(1, 2\frac{1}{2})$ $(4, 2\frac{1}{2})$ $(5, 2\frac{1}{2})$ $(6, 2\frac{1}{2})$

(f) The second number is always $2\frac{1}{2}$.

(g) The dotted line is a line of symmetry.

D3

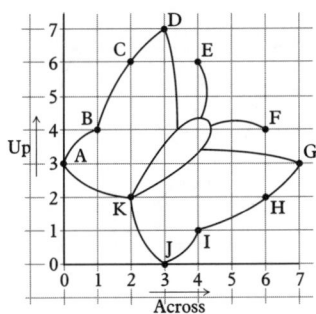

(a) The co-ordinates of the mid-point of DG are $(5, 5)$.

(b)

Point	Partner	Mid-point
A	J	$(1\frac{1}{2}, 1\frac{1}{2})$
B	I	$(2\frac{1}{2}, 2\frac{1}{2})$
C	H	$(4, 4)$
D	G	$(5, 5)$
E	F	$(5, 5)$

(c) The up co-ordinate is equal to the across co-ordinate.

(d) K is its own partner.

E1

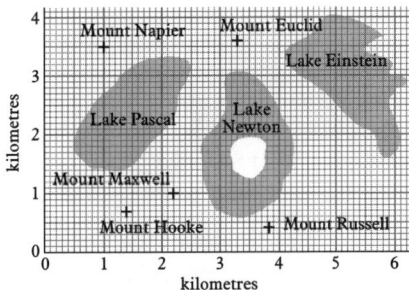

Mount Maxwell $(2\cdot2, 1)$
Mount Russell $(3\cdot8, 0\cdot4)$
Mount Napier $(1, 3\cdot5)$
Mount Euclid $(3\cdot3, 3\cdot6)$

E2

Point	Location
(a) $(2\cdot3, 2\cdot3)$	Land
(b) $(3\cdot1, 1\cdot8)$	Lake Newton
(c) $(1\cdot8, 3\cdot1)$	Lake Pascal
(d) $(3\cdot6, 1\cdot9)$	Land (just)
(e) $(5\cdot2, 3\cdot0)$	Lake Einstein

E3 A point at the middle of Lake Newton is $(3\cdot5, 1\cdot6)$, or nearby.

E4

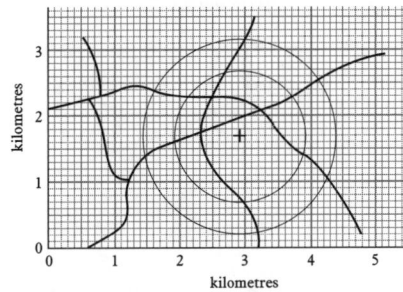

Mr Lamb who is standing at $(2\cdot9, 0\cdot4)$ is in slight danger.

E5

Position	
(a) $(0\cdot4, 2\cdot9)$	Safe
(b) $(3\cdot2, 0\cdot6)$	Slight danger
(c) $(2\cdot3, 2\cdot1)$	Great danger
(d) $(3\cdot7, 0\cdot3)$	Safe
(e) $(3\cdot7, 0\cdot9)$	Slight danger
(f) $(3\cdot7, 2\cdot9)$	Slight danger

E6 The distance across the 'great danger' circle is 2 km. (*We call the distance across a circle its diameter.*)

F1

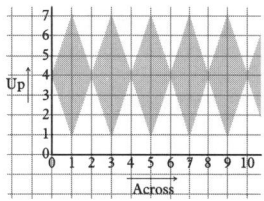

(a) The co-ordinates of the top of the sixth rhombus are (11, 7).

(b) The bottom of the seventh rhombus has co-ordinates (13, 1).

F2

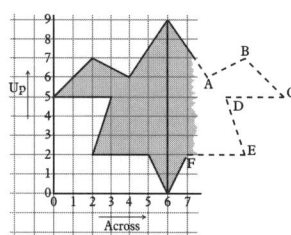

| A (8, 6) | B (10, 7) | C (12, 5) |
| D (9, 5) | E (10, 2) | F (7, 2) |

G1 (a), (d), (e)

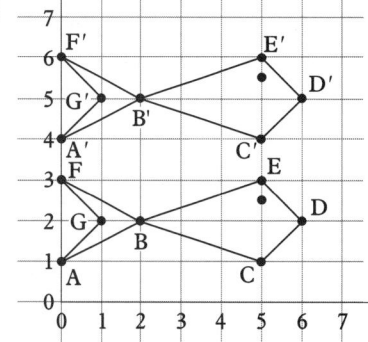

(b), (c)

Point	New point
A (0, 1)	A' (0, 4)
B (2, 2)	B' (2, 5)
C (5, 1)	C' (5, 4)
D (6, 2)	D' (6, 5)
E (5, 3)	E' (5, 6)
F (0, 3)	F' (0, 6)
G (1, 2)	G' (1, 5)

G2 (a)

Point	New point
A (0, 1)	A' (0, 2)
B (2, 2)	B' (4, 4)
C (5, 1)	C' (10, 2)
D (6, 2)	D' (12, 4)
E (5, 3)	E' (10, 6)
F (0, 3)	F' (0, 6)
G (1, 2)	G' (2, 4)

(b)

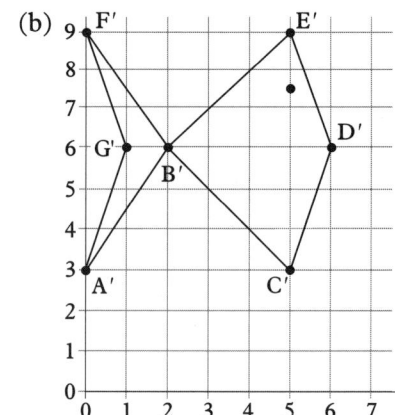

G3 (a)

Point	New point
A (0, 1)	A' (0, 3)
B (2, 2)	B' (2, 6)
C (5, 1)	C' (5, 3)
D (6, 2)	D' (6, 6)
E (5, 3)	E' (5, 9)
F (0, 3)	F' (0, 9)
G (1, 2)	G' (1, 6)

(b)

G4 Show your work to your teacher.

G5 (a), (d)

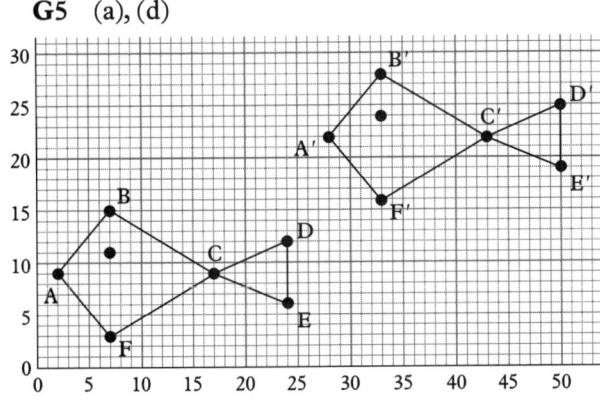

(b), (c)

Point	New point
A (2, 9)	A' (28, 22)
B (7, 15)	B' (33, 28)
C (17, 9)	C' (43, 22)
D (24, 12)	D' (50, 25)
E (24, 6)	E' (50, 19)
F (7, 3)	F' (33, 16)

Coordinates 2

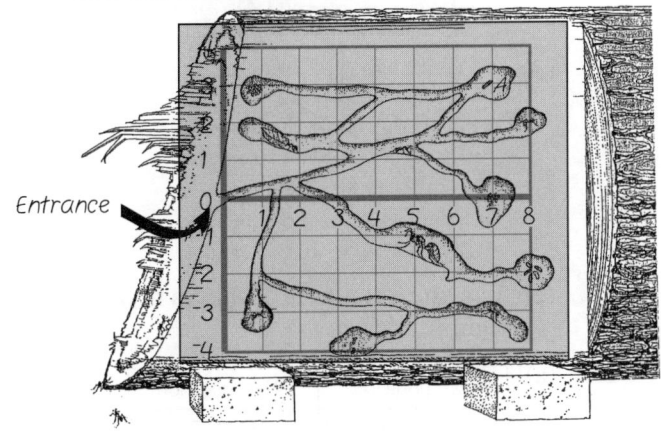

A1 A larva is at (7, 3).

A2 An ant is at (7, ⁻3).

A3 (a) At (1, ⁻3) is an ant.
(b) Eggs are at (7, 0).
(c) Larvae are at $(3\frac{1}{2}, ⁻4)$.
(d) At (1, 3) there are eggs.

A4 All these points are **below** the level of the entrance.
(b) (3, ⁻2) (f) (7, ⁻4)
(g) (5, ⁻1) (h) (1, ⁻4)

A5 (a) The ant carrying a leaf is at $(2, 1\frac{1}{2})$.
(b) (5, ⁻1) is where the queen ant's head is.
(c) The centre of the chamber which has five larvae in it is at (8, ⁻2).

A6 The coordinates of the entrance are (0, 0).

A7 ▲ When an ant enters the nest and goes to feed the larvae at $(3\frac{1}{2}, ⁻4)$, it turns a corner at $(1\frac{1}{2}, \frac{1}{2})$, (1, ⁻2) and (5, ⁻3).

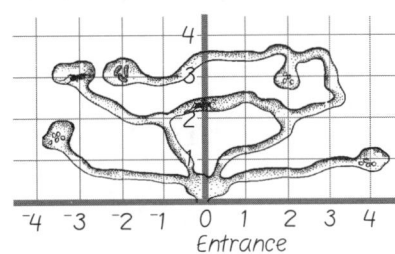

A8 Eggs are at (2, 3).

A9 Larvae are at (⁻2, 3).

A10 (a) An ant is at the point (⁻3, 3).
(b) At $(0, 2\frac{1}{2})$ there is an ant.
(c) Eggs are at the point $(⁻3\frac{1}{2}, 1\frac{1}{2})$.
(d) At the point (4, 1) there are eggs.

B1

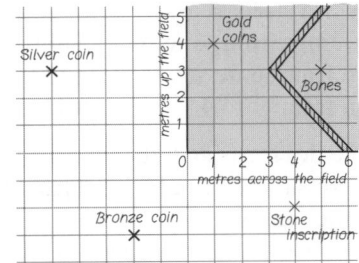

They could have written the places like this using negative coordinates.
(a) Silver coin (⁻5, 3)
(b) Stone inscription (4, ⁻2)
(c) Bronze coin (⁻2, ⁻3)

B2

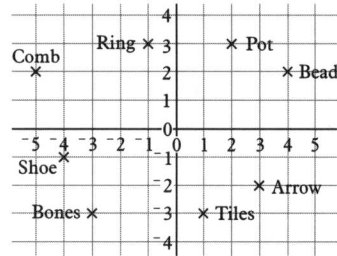

C1 (a) $(^-3,^-2)$ $(^-2,^-1)$ $(^-3,^-1)$ $(2,^-1)$ $(^-1,^-2)$
      S   **P**   **O**   **R**   **T**

    (b) $(^-1,2)$ $(2,^-1)$ $(2,1)$ $(1,^-2)$ $(0,2)$
      D   **R**   **I**   **V**   **E**

C2 IF YOU CAN DECODE THIS YOU ARE DOING WELL.

D1 (a) $^-2 + 3 = 1$ (b) $^-4 + 3 = ^-1$
    (c) $^-2 + 6 = 4$ (d) $^-1 + 1 = 0$
    (e) $^-2 + 2 = 0$ (f) $^-2 + 1 = ^-1$

D2 (a) $2 - 4 = ^-2$ (b) $1 - 3 = ^-2$
    (c) $2 - 5 = ^-3$ (d) $^-1 - 2 = ^-3$
    (e) $^-2 - 3 = ^-5$ (f) $0 - 4 = ^-4$

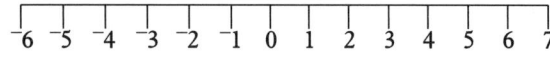

D3 The flea starts at 0.
He goes from 0 to 2,   **Add 2,**
then from 2 to $^-2$,   **subtract 4,**
then from $^-2$ to 1.   **add 3.**

D4 The flea starts at $^-3$.
He jumps from $^-3$ to $^-6$,   **Subtract 3,**
then from $^-6$ to $^-1$,   **add 5,**
then from $^-1$ to 4.   **add 5.**

D5 If the flea starts at 2, then makes these three jumps: subtract 3, add 4, subtract 6, he finishes on $^-3$.
(Does the order of the jumps make any difference to where he ends up?)

D6 Here are the number machines in their correct positions.

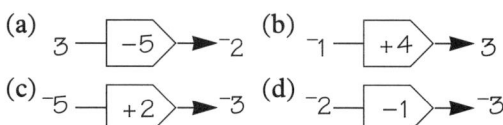

E1 Here are the answers to parts (b) and (c).

Coordinates of corner	New coordinates of corner
A $(3, 5)$	$(^-3, 5)$
B $(5, 4)$	$(^-1, 4)$
C $(3, 1)$	$(^-3, 1)$
D $(1, 4)$	$(^-5, 4)$

(d) The coordinates have changed like this: the across coordinate has had 6 subtracted from it.
(Check the table if you are not quite sure!)

E2

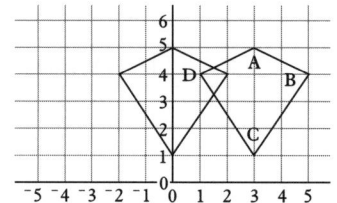

E3 Here are the answers to parts (b) and (c).

Position of corner	New position of corner
A $(^-3, 5)$	$(2, 5)$
B $(^-1, 4)$	$(4, 4)$
C $(^-3, 1)$	$(2, 1)$
D $(^-5, 4)$	$(0, 4)$

(d)

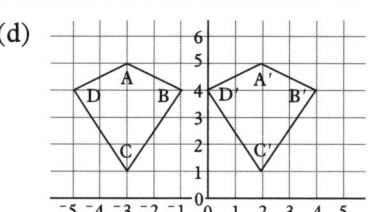

(e) The kite has moved 5 squares to the right.

E4 (a)

Old position	New position
A $(3, 5)$	A' $(3, ^-1)$
B $(5, 4)$	B' $(5, ^-2)$
C $(3, 1)$	C' $(3, ^-5)$
D $(1, 4)$	D' $(1, ^-2)$

(b) Here is the machine for the slide.

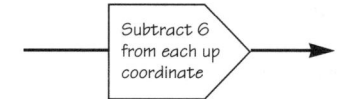

E5

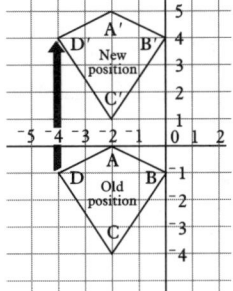

The machine for this slide is:

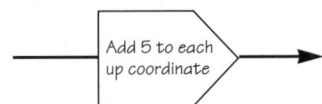

E6

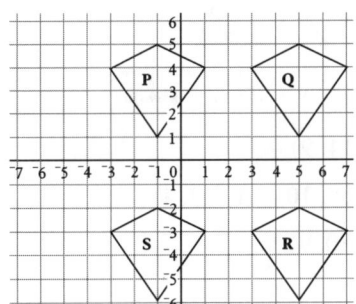

(a) The rule to slide the kite from Q to R is:

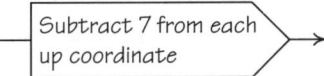

(b) To slide from R to S the rule is:

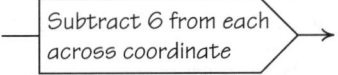

(c) This rule will slide the kite from P to S:

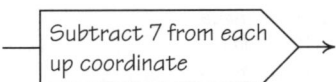

(d) This is the rule to slide from S to P:

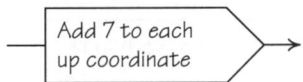

(e) To get from Q to P in two moves the rule is:
It does not matter in which order you make the moves.

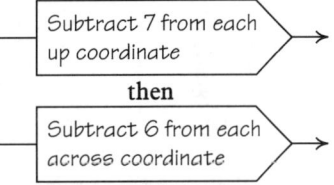

then

(f) This pair of rules will slide the kite from P to R:

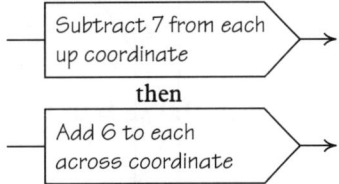

then

(g) These will slide the kite from R to P:

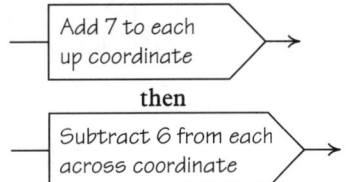

then

F1

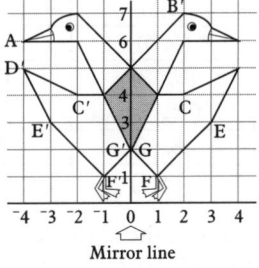

Mirror line

Point		Image	
A	(⁻4, 6)	A'	(4, 6)
B	(⁻2, 7)	B'	(2, 7)
C	(2, 4)	C'	(⁻2, 4)
D	(4, 5)	D'	(⁻4, 5)
E	(3, 3)	E'	(⁻3, 3)
F	(1, 1)	F'	(⁻1, 1)
G	(0, 2)	G'	(0, 2)

13

F2 Here are the points and their images.

(b)

Point	Image
A ($^-$3, 0)	A' ($^-$3, 0)
B ($^-$3, 3)	B' ($^-$3, $^-$3)
C ($^-$1, 3)	C' ($^-$1, $^-$3)
D ($^-$1, 0)	D' ($^-$1, 0)
E (1, $^-$3)	E' (1, 3)
F (3, $^-$3)	F' (3, 3)
G (1, 0)	G' (1, 0)

Here is the shape and its image.

(a), (c)

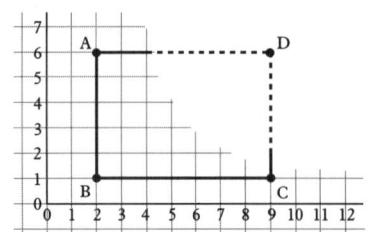

F3 (a)

Point	Coordinates	Coordinates of image
A	(4, 2)	(2, 4)
B	($^-$1, 1)	(1, $^-$1)
C	(2, 1)	(1, 2)
D	(0, 3)	(3, 0)
E	(0, $^-$2)	($^-$2, 0)
F	($^-$5, $^-$3)	($^-$3, $^-$5)
G	(6, $^-$1)	($^-$1, 6)
H	($^-$6, 4)	(4, $^-$6)

(b) For each point the across co-ordinate changes places with the up co-ordinate.

Co-ordinate patterns

A1

(a) The co-ordinates of D are (9, 6).
(b) The co-ordinates of S are (7, 3).
(c) N (3, 6)

A2

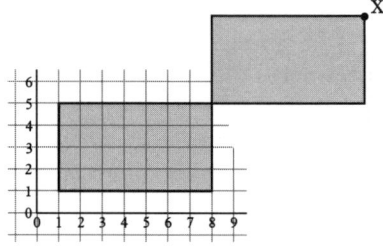

(a) X (15, 9)
(b) ($11\frac{1}{2}$, 7) are the co-ordinates of the centre of the upper rectangle.

A3

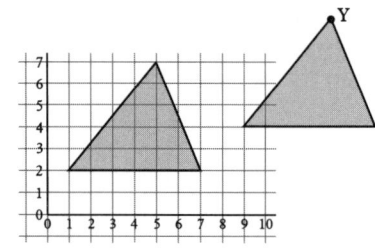

Y (14, 9)

A4 X (10, 10)   Y (13, 4)

A5 D (10, 9)

B1

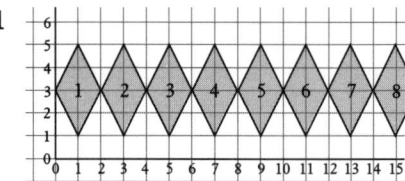

(a) The co-ordinates of the tops of the ninth, tenth and eleventh rhombuses are (17, 5), (19, 5) and (21, 5).

Right hand corner of the . . .	Co-ordinates
(b) 8th rhombus	(16, 3)
(c) 9th rhombus	(18, 3)
(d) 10th rhombus	(20, 3)
(e) 11th rhombus	(22, 3)
(f) 20th rhombus	(40, 3)
(g) 53rd rhombus	(106, 3)

Can you see the pattern in the across co-ordinates? Look at this table again if you can't.

(h) The co-ordinates of the centre of the 53rd rhombus are (105, 3).

(i) The centre of the 68th rhombus has co-ordinates (135, 3).

(j) The **bottom** of the 100th rhombus is the point (199, 1).

B2 ▲

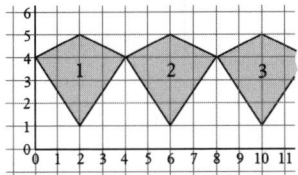

(a) (198, 5) are the co-ordinates of the top of the 50th kite.

(b) The point (100, 1) is not at the bottom of a kite. (How can you tell? Could you convince someone that this is true?)

(c) These points are at the bottom of kites:
(202, 1) (198, 1) (58, 1)
(How did you decide this?)

B3

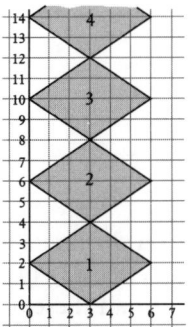

Top of the . . .	Co-ordinates
(a) 4th rhombus	(3, 16)
(b) 5th rhombus	(3, 20)
(c) 7th rhombus	(3, 28)
(d) 17th rhombus	(3, 68)

Can you see a connection between the number of the rhombus and its up co-ordinate?

(e) The co-ordinates of the **right-hand corner** of the 83rd rhombus is (6, 330).

B4

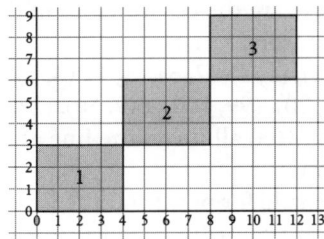

Top right-hand corner of . . .	Co-ordinates
(a) 4th rectangle	(16, 12)
(b) 5th rectangle	(20, 15)
(c) 6th rectangle	(24, 18)
(d) 10th rectangle	(40, 30)
(e) 30th rectangle	(120, 90)

(f) (100, 78) are the co-ordinates of the top left-hand corner of the 26th rectangle.

(g) The centre of the 100th rectangle is $(398, 298\frac{1}{2})$.

B5 ▲ The four corners of the 24th parallelogram in the pattern are (144, 96), (142, 93), (138, 92), (140, 95).

B6 (a) (22, 0), (24, 5), (20, 5)
(b) (62, 0), (64, 5), (60, 5)
(c) (98, 10), (96, 5), (100, 5)
(d) (102, 0), (100, 5), (104, 5)

B7

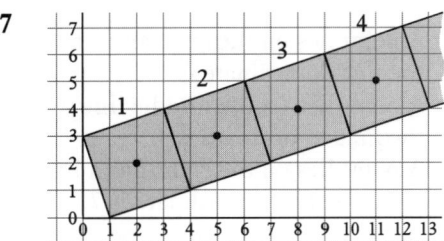

(a) The centre of the 10th square is at (29, 11).

(b) (32, 12) is the centre of the 11th square.

(c) The 20th square has its centre at (59, 21).

B8

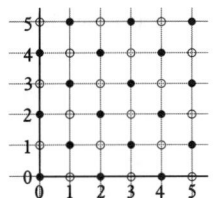

(a) (7, 0) is ○   (b) (0, 10) is ●
(c) (20, 7) is ○   (d) (30, 12) is ●
(e) There is ● at (19, 23).
(f) There is ○ at (71, 14).
(g) (31, 5) is ●   (h) (23, 23) is ●

B9 These are found at each point:
(a) ● (8, 1)   (b) ○ (0, 12)
(c) ○ (8, 10)   (d) × (13, 11)
(e) ● (7, 16)   (f) ● (15, 18)
(g) ● (20, 19)   (h) ○ (22, 22)

C1

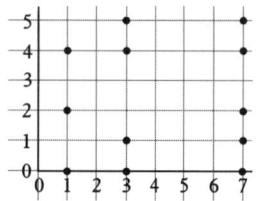

(a)

(A) On left-hand line	(B) On middle line	(C) On right-hand line	(D) Not on any line
(1, 19)	(3, 6)	(7, 19)	(2, 7)
(1, 6)	(3, 29)	(7, 38)	(19, 1)
(1, 80)	(3, 32)	(7, 3)	(5, 1)
		(7, 7)	(6, 38)
			(76, 1)
			(100, 7)
			(0, 3)
			(0, 1)
			(33, 46)
			(2, 13)
			(13, 3)

(b) A point lies on the left-hand line when the first co-ordinate is 1.

(c) A point lies on the middle line when the first co-ordinate is 3.

(d) A point lies on the right-hand line when the first co-ordinate is 7.

C2 (a) The rule for points on the middle line is first co-ordinate = 3.

(b) The rule for points on the right-hand line is first co-ordinate = 7.

C3 All points like (4, 1), (4, 2), (4, 3) and so on fit the rule first co-ordinate = 4.

C4 All these points lie on the line with the rule first co-ordinate = 8:
(8, 8), (8, 13), (8, 10).

C5

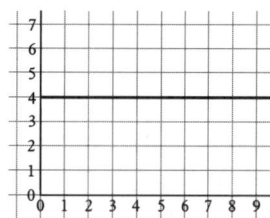

(a) All these points lie on the line:
(2, 4), (3, 4), (28, 4), (93, 4), (6·8, 4), (4, 4), (0, 4).

(b) The rule for the points on the line is second co-ordinate = 4.

C6

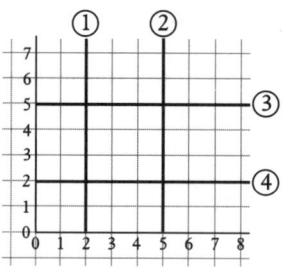

(a) Line 3 has the rule second co-ordinate = 5.

(b) Line 1, first co-ordinate = 2.
Line 2, first co-ordinate = 5.
Line 4, second co-ordinate = 2.

C7 The line with the rule first co-ordinate = 6 crosses the line with the rule second co-ordinate = 3 at the point (6, 3).

C8

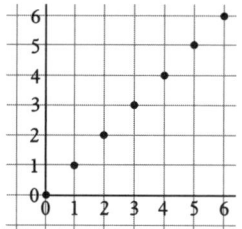

(a) These points lie on the line: (10, 10), (0, 0), (4, 4), (7, 7), (83, 83), (47, 47).

(b) The rule which fits all these points is the first co-ordinate is equal to the second co-ordinate.

C9

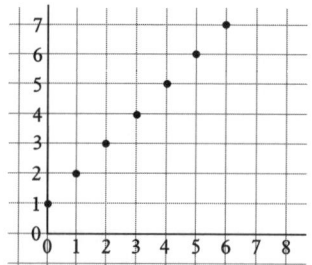

(a) These are the co-ordinates of the first six points:
(0, 1), (1, 2), (2, 3), (3, 4), (4, 5), (5, 6).

(b) (19, 20), (99, 100), (49, 50), (28, 29), (162, 163) all lie on the line of marked points.

(c) The rule for points on the line is the second co-ordinate is 1 greater than the first co-ordinate.

D1 (a) 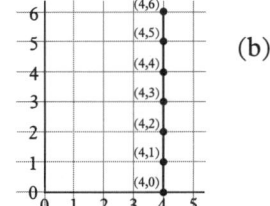 (b)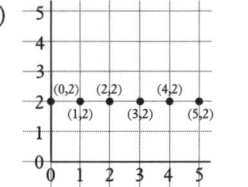

(a) The rule for this line is $x = 4$.
(b) This line has the rule $y = 2$.

D2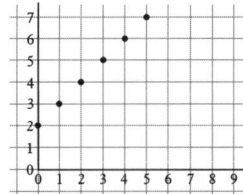

(a) The co-ordinates of the points are:
$(0, 2), (1, 3), (2, 4), (3, 5), (4, 6), (5, 7)$.
(b) These points are all on the line of marked points.
$(7, 9), (10, 12), (16, 18), (19, 21),$
$(27, 29), (100, 102)$
(c) In x, y language the rule for points on the line can be any of these:
$y = x + 2$, or $y - 2 = x$, or $y - x = 2$.

D3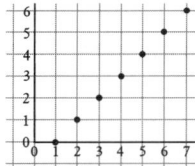

(a) $(1, 0), (2, 1), (3, 2), (4, 3), (5, 4), (6, 5),$
$(7, 6)$ are the co-ordinates of the marked points.
(b) The points $(10, 9), (26, 25), (38, 37)$ are all on the line.
(c) In x, y language the rule for points on the line can be any of these.
$y = x - 1$ or $y + 1 = x$ or $x - y = 1$

D4 (a) The co-ordinates of the marked points are: $(0, 0), (1, 2), (2, 4), (3, 6), (4, 8)$.
(b) These points are on the line.
$(100, 200), (10, 20), (31, 62), (73, 146)$
(c) In x, y language the rule for points on the line can be either of these.
$y = 2 \times x$ or $y \div 2 = x$

D5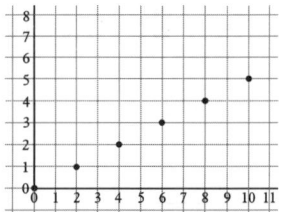

(a) The rule for this line is $y = x \div 2$ or $2 \times y = x$.
(b) All these points are on the line.
$(18, \mathbf{9}), (48, \mathbf{24}), (\mathbf{66}, 33), (\mathbf{58}, 29),$
$(43, \mathbf{21\frac{1}{2}}), (71, \mathbf{35\frac{1}{2}})$.

D6 (a) The points $(2, 5), (0, 1), (4, 9)$ all fit the rule $(x \times 2) + 1 = y$.
(b) Your points should be in a straight line.
(c) All these points fit the rule:
$(1, \mathbf{3}), (5, \mathbf{11}), (10, \mathbf{21}), (13, \mathbf{27}), (\mathbf{9}, 19),$
$(\mathbf{16}, 33), (\mathbf{50}, 101), (3\cdot5, \mathbf{8}), (5, 11)$.

D7 (a) $(0, 2), (1, 5), (2, 8)$, and so on, all fit the rule $(x \times 3) + 2 = y$.
See your teacher if you are unsure.
(b) These points: $(23, 71), (0\cdot5, 3\cdot5),$
$(200, 602)$, and $(2\cdot8, 10\cdot4)$ are all on the line.
(c) These points are all on the line:
$(6, \mathbf{20}), (21, \mathbf{65}), (\mathbf{8}, 26), (\mathbf{33}, 101),$
$(\mathbf{2\cdot5}, 9\cdot5)$.

D8 These points fit the rule $(x \times 3) - 1 = y$:
$(0, {}^-\mathbf{1}), (4, \mathbf{11}), (2, 5), (7, \mathbf{20}), (15, \mathbf{44})$.

D9 (a) $(x \times 2) + 1 = y$ is the only rule which fits $(2, 5)$ and $(3, 7)$.
(b) $(x \times 4) - 3 = y$ fits both the points $(1, 1)$ and $(2, 5)$.
(c) $(x \times 3) - 2 = y$ fits the points $(1, 1)$ and $(3, 7)$.

D10 (a) The rule which fits all these points: $(1, 5), (3, 9), (5, 13), (20, 43), (30, 63)$ and $(100, 203)$ is $(x \times 2) + 3 = y$.
(b) These points also fit the rule:
$(0, \mathbf{3}), (1, \mathbf{5}), (2, \mathbf{7}), (3, \mathbf{9}), (4, \mathbf{11}), (5, 13)$.
(c) Your points should be in a straight line.

D11 (a) The rule which fits all these points: $(1, 1), (4, 10), (6, 16), (10, 28), (50, 148),$ $(100, 298)$ is $(x \times 3) - 2 = y$.
(b) Your points should be in a straight line.

Coordinates 2: extension

A1 R($^-$6, $^-$3), S($^-$6, $^-$2)

A2
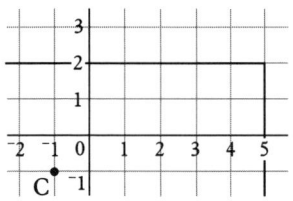

(*Remember, the centre of the rectangle is at* ($^-$1, $^-$1).)

The coordinates of the other three corners are ($^-$7, 2), ($^-$7, $^-$4), and (5, $^-$4).

A3
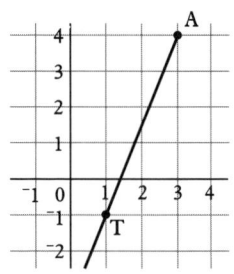

(a) If T is halfway along the line from A to B, then B is at ($^-$1, $^-$6).

(b) If T is one-third of the way from A to a point C, then C is at ($^-$3, $^-$11).

A4
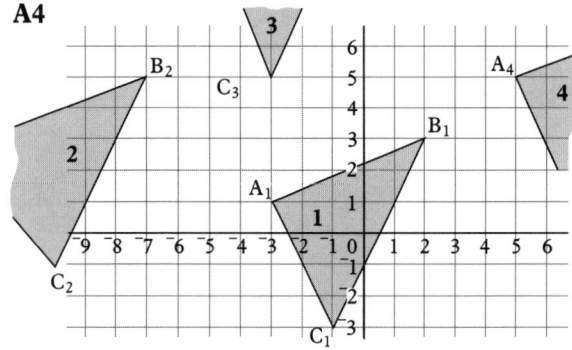

(a) A_2 ($^-$12, 3)   (b) A_3 ($^-$5, 9)

(c) B_3 (0, 11)   (d) B_4 (10, 7)

(e) C_4 (7, 1)

A5 (a) Q (2, 1), R (3, $^-$2)

(b) Q ($^-$4, 3), R ($^-$3, 0)

B1
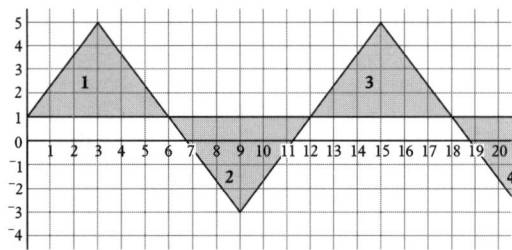

(a) The corners of the fourth triangle are at (18, 1), (21, $^-$3) and (24, 1).

(b) The top corner of triangle 5 is at (27, 5).

(c) The bottom corner of triangle 8 is at (45, $^-$3).

B2
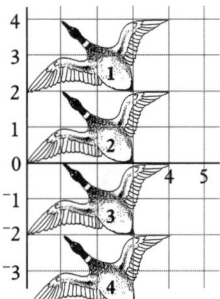

(a) The right-hand wing-tip of the 10th duck is at (4, $^-$14).

(b) Its left-hand wing-tip is at (0, $^-$16).

(c) The tail of the 15th duck is at (3, $^-$26).

(d) The 50th duck has its tail at (3, $^-$96).

(e) The tip of the bill of the 50th duck is at the point (1, $^-$94).

(*It is 2 to the left and 2 above its tail!*)

B3
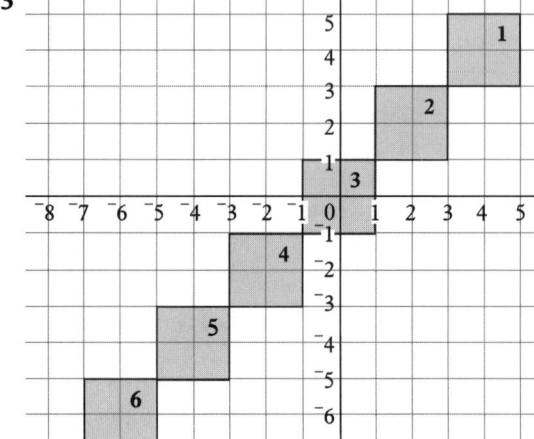

18

(a) The centre of the 10th square is at (⁻14, ⁻14).

(b) The centre of the 15th square is at (⁻24, ⁻24).

(c) The 40th is at (⁻74, ⁻74).

(d) The top left-hand corner of the 40th square is at (⁻75, ⁻73).

(e) The bottom right-hand corner of the 100th square is at (⁻193, ⁻195).

B4

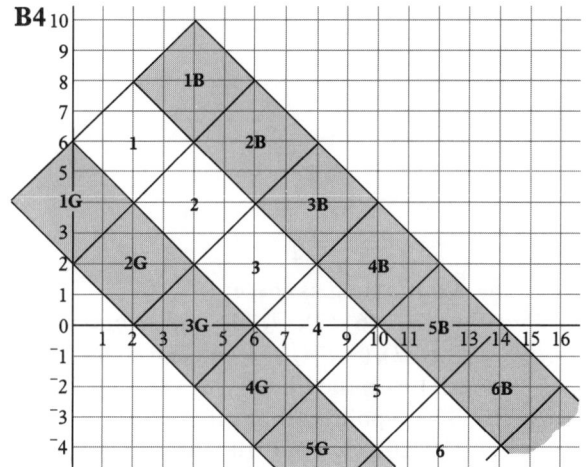

(a) The top corner of square 6 is (12, ⁻2) and of (b) square 7 is (14, ⁻4).

(c) Square 8 (16, ⁻6)

(d) square 40 (80, ⁻70).

(e) The right-hand corner of square 20 is at (42, ⁻32).

(f) The right-hand corner of square 20B is at (44, ⁻30).

(g) The top corner of 21B is at (44, ⁻30).

(h) The left-hand corner of square 25 is (48, ⁻42).

(i) The left-hand corner of square 25G is (46, ⁻44).

(j) (50, ⁻48) is the bottom corner of square 26G.

C1

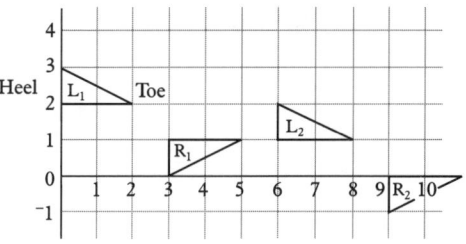

(a) The three corners of L_3 are at (12, 0), (12, 1) and (14, 0).

(b) The three corners of R_3 are at (15, ⁻1), (15, ⁻2) and (17, ⁻1).

C2 The toe of footprint L_5 is at (26, ⁻2).

C3 The heel points of R_5 are at (27, ⁻3) and (27, ⁻4).

C4 The point (21, ⁻2) is part of a right footprint.

C5 The point (20, 0) is not part of a right or left footprint.

C6 This table gives the positions of Lurch's right foot.
(Look for the patterns in the numbers.)

R_1	(3, 0)	(3, 1)	(5, 1)
R_2	(9, ⁻1)	(9, 0)	(11, 0)
R_3	(15, ⁻2)	(15, ⁻1)	(17, ⁻1)
R_{50}	(297, ⁻49)	(297, ⁻48)	(299, ⁻48)
R_{60}	(357, ⁻59)	(357, ⁻58)	(359, ⁻58)

C7 After 89 jumps he will be standing on the up-axis.

D1 (a)

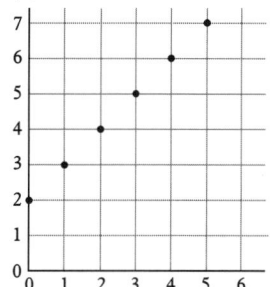

x	y
0	2
3	5
6	8
10	12
23	25
29	31

(b) These points lie on the line: (45, 47) and (99, 101).

(c) These rules fit all the points.
$$x + 2 = y, \quad y - x = 2, \quad y = 2 + x$$
$$y = x + 2, \quad y - 2 = x, \quad x = y - 2$$

19

D2

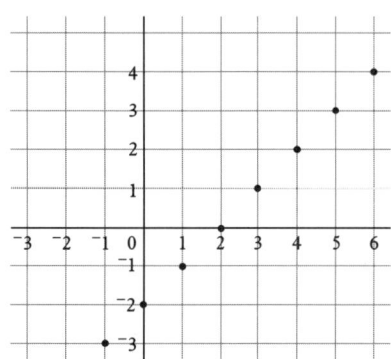

(a) All these points lie on the line.

x	y
2	5
⁻3	0
⁻5	⁻2
⁻8	⁻5
14	17

(b) Any of the following give a rule for points on the line.
$x + 3 = y$, $y = x + 3$, $3 + x = y$,
$y = 3 + x$, $y - x = 3$, $3 = y - x$,
$y - 3 = x$, $x = y - 3$

(c) These points lie on the line: (20, 23), (⁻23, ⁻20).

(d) All these points are on the line.
(17, **20**), (⁻11, ⁻**8**), (**11**, 14),
(**18**, 21), (⁻18, ⁻**15**), (⁻**21**, ⁻18),
(⁻**10**, ⁻7), (⁻**68**, ⁻**65**)

D3

(a) All these points lie on the line.

x	y
3	1
6	4
0	⁻2
⁻1	⁻3
12	10
⁻5	⁻7
⁻20	⁻22

(b) Any of these rules fit points on the line.
$y = x - 2$, $x - 2 = y$, $y + 2 = x$,
$x = y + 2$, $2 + y = x$, $x = 2 + y$,
$x - y = 2$, $2 = x - y$

(c) These two points lie on the line: (16, 14), (⁻18, ⁻20).

D4
▲

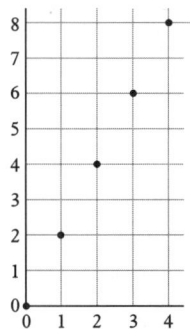

(a) Points such as these:
(5, 10), (6, 12), (7, 14), and so on;
(⁻1, ⁻2), (⁻2, ⁻4), (⁻3, ⁻6), and so on;
all lie on the line.

(b) These points all lie on the line.
(10, **20**), (5, 10), (⁻**10**, ⁻20), (15, **30**),
(**22**, 44), (⁻**8**, ⁻16)

(c) Any of the following rules fit points on the line.
$y = 2 \times x$, $2 \times x = y$, $x = \frac{1}{2} \times y$,
$\frac{1}{2} \times y = x$

D5 (a) Any of these:
$y = x + 1$, $x + 1 = y$, $1 + x = y$,
$y = 1 + x$, $x = y - 1$, $y - 1 = x$,
$y - x = 1$, $1 = y - x$

(b) Rules for lines parallel to $y = x + 1$
are $y = 3 + x$, $y = x + 2$, $y = x + 3$,
$y = x + 4$, and so on
or $y = x - 1$, $y = x - 2$, $y = x - 3$, and so on.
If you have written your rule in a different way and think it is right, show it to your teacher.

D6

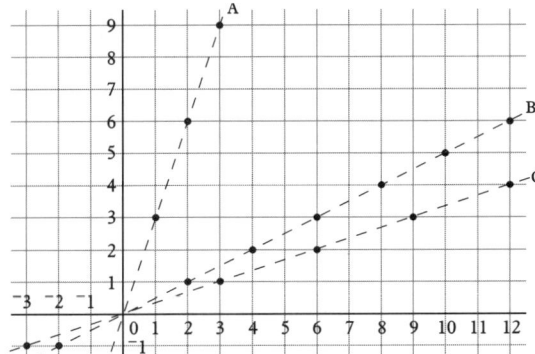

(a) **Line A** (⁻5, ⁻15), (20, 60), (⁻3, ⁻9),
    (100, 300)
   Line B (12, 6), (⁻18, ⁻9), (⁻12, ⁻6)
   Line C (⁻15, ⁻5), (⁻12, ⁻4), (75, 25)
    (⁻12, ⁻4), (⁻300, ⁻100)

(b) Rules for the three lines are:
   Line A $y = 3 \times x$ or $x = \frac{1}{3} \times y$
   Line B $y = \frac{1}{2} \times x$ or $x = 2 \times y$
   Line C $y = \frac{1}{3} \times x$ or $x = 3 \times y$

E1

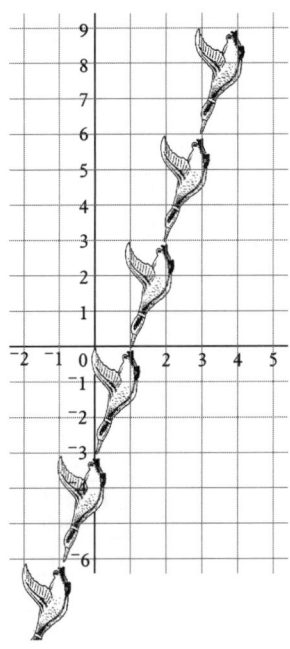

(a) Either of these rules fits the line
   through all the ducks' wing-tips.
   $y = 3 \times x$ or $x = \frac{1}{3} \times y$

(b) Either of these rules fits the line
   through all the ducks' bills.
   $y = (3 \times x) - 3$ or $x = (\frac{1}{3} \times y) + 1$
   You may have written the rules in a
   different way, show your teacher if you
   have.

E2

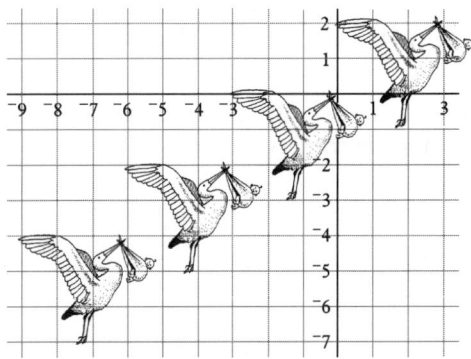

(a) These rules give the line which goes
   through the tips of the storks' bills.
   $2 \times x = 3 \times y$ or $y = \frac{2}{3} \times x$
   or $x = 1\frac{1}{2} \times y$

(b) These rules fit the line through the
   storks' wing-tips.
   $y = (\frac{2}{3} \times x) + 2$ or $3 \times y = (2 \times x) + 6$
   or $x = (1\frac{1}{2} \times y) - 3$

(c) The two lines are parallel.
   You may have written your rules in a
   different way, if so show your teacher.

F1 (a) They will touch after 3 minutes.
   (b) (0, 2)   (c) $y = x + 1$

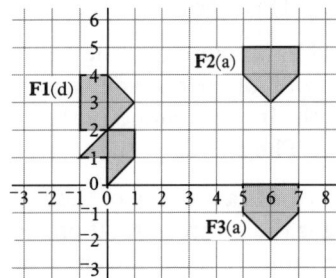

F2 (b) $y = 3$ (c) (6, 3)

F3 (b) (6, ⁻2)

F4 (a) No
   (b) $x + y = 3$ or $y = 3 - x$ or $x = 3 - y$

21

Straight line graphs

A1 The crosses show all the possible positions for the corners of a square.

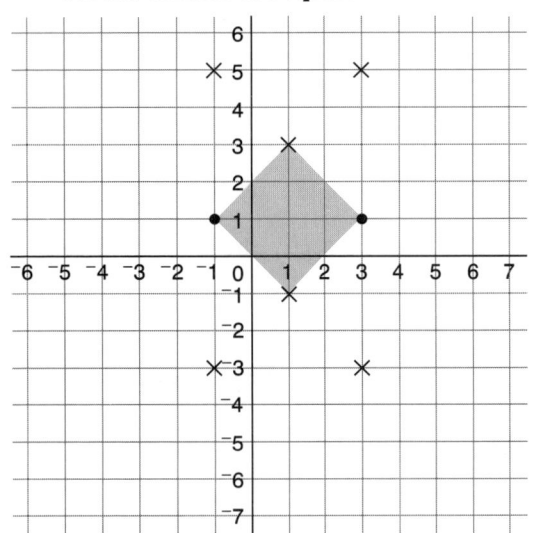

A2 If you look at their coordinates you will see that they have written down their coordinates in the wrong order. For example $(5, {}^-1)$ rather than $({}^-1, 5)$.

A3 You should have found that the new shape was a reflection of the one you started with. Was the line of symmetry the same every time? The area stays the same.

A4 The new shape is a rotation of the first one. What is the centre of rotation? Is it always the same? What is the angle of rotation?

B1

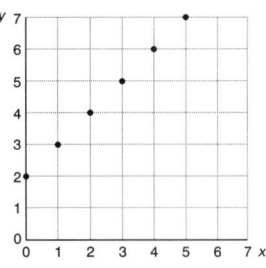

(a) The table shows the coordinates of some of the points on this line.

x (across coordinate)	y (up coordinate)
0	2
3	5
5	7
10	12
18	20

(b)

Coordinate	Is it on the line?
(37, 35)	No
(35, 37)	Yes
(100, 101)	No
(98, 100)	Yes
(101, 99)	No

If you look at the points in the table, which are on the graph you will see that the up coordinate (y) is always two more than the across co-ordinate (x).

(c) One way to check if a rule fits a point, is to put in the values of x and y to see if they fit the rule.

Point	Rule	Left-hand side	Right-hand side	Does it fit?
(3, 5)	$x + 2 = y$	$3 + 2 = 5$	5	Yes
(3, 5)	$y - 2 = x$	$5 - 2 = 3$	3	Yes
(3, 5)	$x + y = 2$	$3 + 5 = 8$	2	No
(3, 5)	$y = x + 2$	5	$3 + 2 = 5$	Yes
(3, 5)	$x = y + 2$	3	$5 + 2 = 7$	No
(3, 5)	$x = y - 2$	3	$5 - 2 = 3$	Yes

If a rule fits one point, can we be sure it will fit all the points?

B2

Point	Rule	Left-hand side	Right-hand side	Does it fit?
(1, 4)	$x + y = 4$	$1 + 4 = 5$	4	No
(1, 4)	$x - 3 = y$	$1 - 3 = {}^-2$	4	No
(1, 4)	$y = 4x$ $[4x \text{ is } 4 \times x]$	4	$4 \times 1 = 4$	Yes
(1, 4)	$y = 4$	4	4	Yes
(1, 4)	$y - x = 1$	$4 - 1 = 3$	1	No
(1, 4)	$y = x + 3$	4	$1 + 3 = 4$	Yes

B3 ▲

Point	Rule	Left-hand side	Right-hand side	Does it fit?
$(1, 4)$	$y = x + 4$	4	$1 + 4 = 5$	No
$(1, 5)$	$y = x + 4$	5	$1 + 4 = 5$	Yes
$(0, 4)$	$y = x + 4$	4	$0 + 4 = 4$	Yes
$(3, 7)$	$y = x + 4$	7	$3 + 4 = 7$	Yes
$(5, 1)$	$y = x + 4$	1	$5 + 4 = 9$	No

B4 Which of these points fit the rule $x = 1$?
(It only says that $x = 1$: it does not mention
the value of y, it can take any value.)
$(1, 1)$, yes $(2, 1)$, no $x = 2$
$(1, 2)$, yes $(1, 0)$, yes

B5

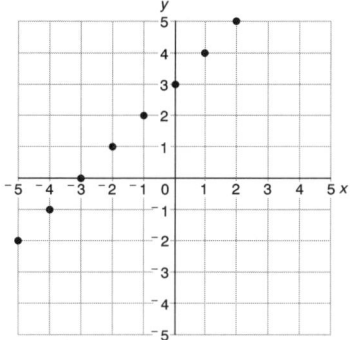

(a) This table shows some points which
are on the line of points.

x	y
2	5
⁻2	1
1	4
⁻3	0
⁻8	⁻5
14	17
0	3

(b) Looking at the table you can see that
the value of y for a point is 3 more
than the value of x for the same point.
We can write this as $y = x + 3$
or $x + 3 = y$ (or even $y - x = 3$).

(c) Your own checks to see your rule
works.

(d) These points fit the rule $y = x + 3$ or
$x + 3 = y$ (or $y - x = 3$):
$(20, 23)$, $(⁻10, ⁻7)$
but these points don't:
$(23, 20)$, $(⁻7, ⁻10)$.

(e) These points are all on the line
$y = x + 3$.
$(17, \mathbf{20})$, $(\mathbf{10}, 13)$, $(18, \mathbf{21})$,
$(⁻8, ⁻5)$, $(⁻\mathbf{10}, ⁻7)$

B6

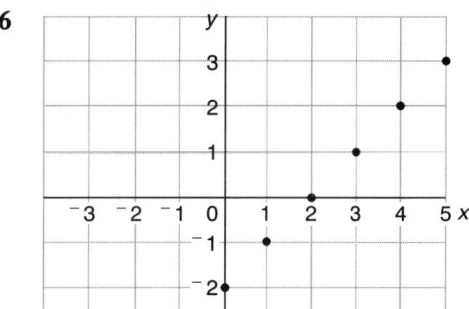

(a) Here is the completed table for this
line of points.

x	y
3	1
4	2
0	⁻2
⁻1	⁻3
2	0
⁻3	⁻5

(b) A rule which fits the line of points is
$y = x - 2$ (or $x - 2 = y$).

B7 ▲

Frank is wrong, $x = 2$ works for the point
$(2, 2)$ but not the rest. A rule which does fit
all the points is $y = 2$, check it yourselves.

23

B8 The rule for these points is $x = 3$. Check it for yourselves if you are not sure.

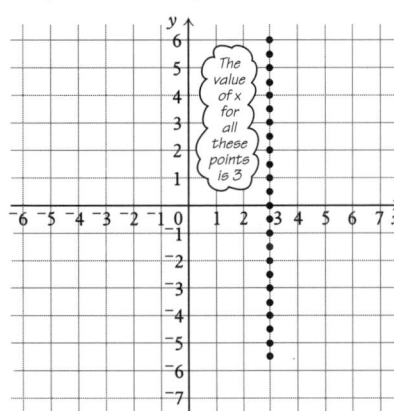

The value of x for all these points is 3

B9 ▲

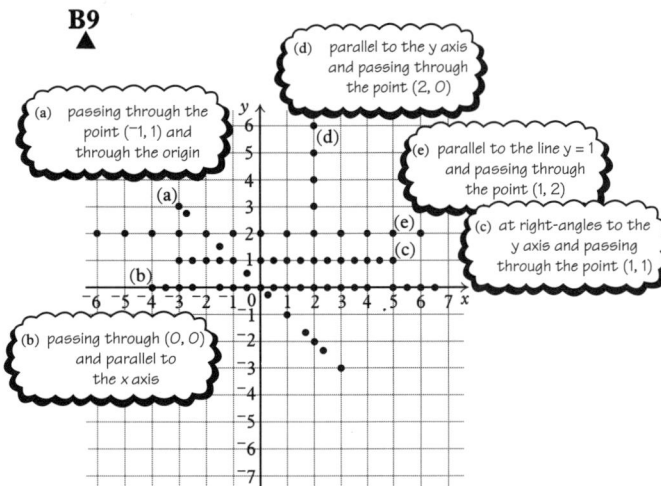

(a) passing through the point (⁻1, 1) and through the origin

(b) passing through (0, 0) and parallel to the x axis

(d) parallel to the y axis and passing through the point (2, 0)

(e) parallel to the line y = 1 and passing through the point (1, 2)

(c) at right-angles to the y axis and passing through the point (1, 1)

B10 (a)

x	0	1	2	3	4	5	6	7	8
y	5	5	5	5	5	5	5	5	5

The rule is $y = 5$.

(b)

x	4	4	4	4	4	4	4	4
y	0	1	2	3	4	5	6	7

The rule is $x = 4$.

Investigate

You may have found it helpful to draw up a table showing $2x$ and $x + 2$ for different values of x. What happens when $x = 0$, $x = 1$, $x = 0.5$, $x = ⁻1$, etc?

B11

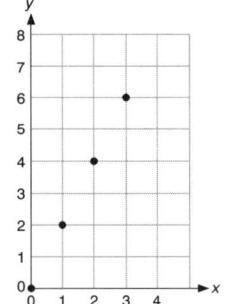

(a) Here are four points which are on the line – you may have found some different ones of your own.
(1·5, 3), (4, 8), (6, 12), (0·5, 1),
(points don't have to be whole numbers!)

(b) All these points lie on the line:
(10, **20**), (5, 10), (15, **30**), (22, 44).

(c) The rule $y = 2x$ fits all the points on the line.

B12

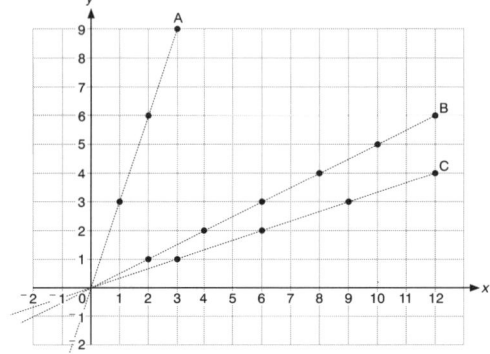

Line	Points on the line	Rule for line
A	(0, 0), (10, 30), (30, 90)	$y = 3x$
B	(0, 0), (12, 6), (14, 7), (20, 10), (10, 5), (100, 50), (60, 30)	$y = x ÷ 2$ or $y = 0.5x$
C	(0, 0), (12, 4) (120, 40), (30, 10), (60, 20)	$y = x ÷ 3$

24

C1

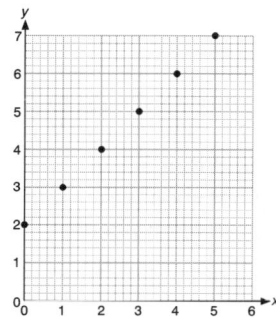

(a) The rule for these points is $y = x + 2$.
    (Check it if you are not sure.)

(b) All these points lie on the line of
    points shown.
    $(0{\cdot}5, 2{\cdot}5)$ $(3{\cdot}5, 5{\cdot}5)$ $(1{\cdot}5, 3{\cdot}5)$ $(4{\cdot}5, 7{\cdot}5)$
    $(2{\cdot}5, 4{\cdot}5)$ $[4{\cdot}5 - 2 = 2{\cdot}5]$

(c) They should lie on the same straight
    line!

(d) You would probably find it hard to tell
    the point $(1{\cdot}01, 3{\cdot}01)$ from $(1{\cdot}02, 3{\cdot}02)$.

C2

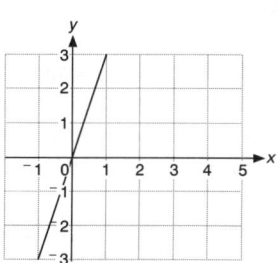

(a) $y = 3x$

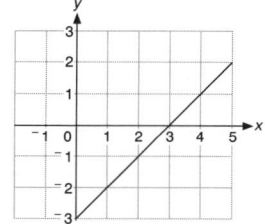

(b) $y = x - 3$

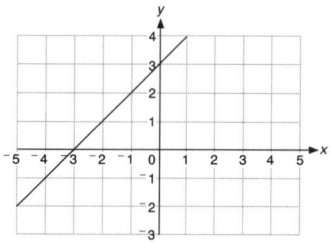

(c) $y = x + 3$

D1 If you look at where the lines cut the **axes**
you can work out which equation fits the
line. If you are not quite sure look back at
some of the equations and lines in Section C.

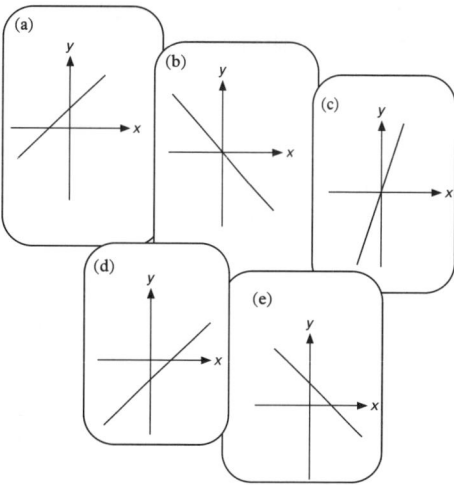

(a) **matches (A)** $y = x + 4$ (the value of y is
    greater than the value of x, $y = 0$
    when $x = {}^-4$.)

(b) **none of the equations** ($y = 0$ when
    $x = 0$, when x is positive, y is negative.)

(c) **matches (B)** $y = 4x$ ($y = 0$ when $x = 0$,
    when x is positive, y is positive.)

(d) **matches (C)** $y = x - 4$, how could you
    tell?

(e) **none of the equations,** how could you
    tell?

D2

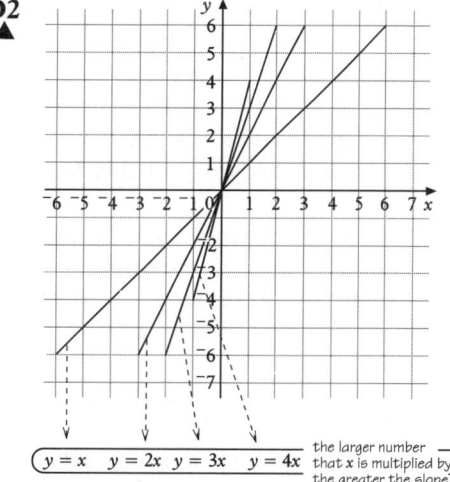

$y = x$ $y = 2x$ $y = 3x$ $y = 4x$ the larger number
that x is multiplied by
the greater the slope

25

D3 If you draw (or plot) the straight line graphs $y = 2x + 1$ and $y = 3x - 1$, the second straight line will be the steeper. What else is different about the two graphs? *Can you tell by looking at them which one of these is the steepest?*
$y = x$ $y = 3x + 3$ $y = 4x - 10$
$y = 4.5x - 6.5$
Experiment for yourselves.

E1 ▲ (a) Here are some different kinds of straight line graph.

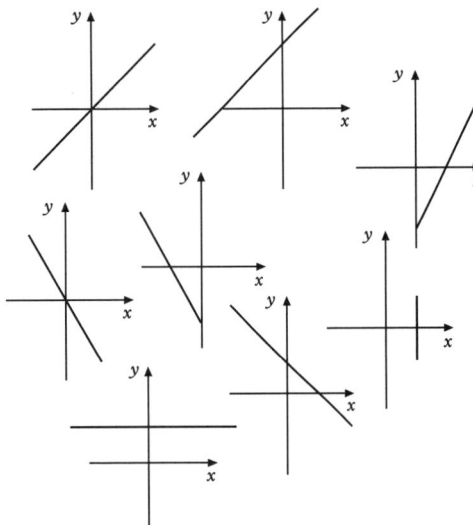

(b) This is an equation of an equation for a straight line graph!
$y = ax + b$, the letters a and b stand for numbers.
In the straight line graph
$y = 3x - 6$, $a = 3$ and $b = {}^-6$,
and $a = 1$ and $b = 4$ in $y = x + 4$.
Here are some rules, you probably found more.

- If b is zero (for example, $y = 7x$) then the line goes through the **origin**.
- If a is zero (for example, $y = 4$) then the straight line is parallel to the **x axis** (page 6).
- The bigger the value of a, the steeper the straight line.
  (What happens when a is negative?)
- What else did you find out? – show your teacher.

E2 ▲ What happens with graphs like $y = x \times x$ or $y = x \times x - 2x$ ($y = x^2 - 2x$)?

E3 ▲ From the graph you should find that $2 \times {}^-2 = {}^-4$, $2 \times {}^-3 = {}^-6$, etc. You should also have found that $5 \times {}^-4 = {}^-20$. What did you find out about other multiplications involving negative numbers?

E4 ▲ All these pairs of straight line graphs are parallel.
Can you spot a rule? (**Hint.** Look at a in $y = ax + b$)

This graph		this graph
$y = 2x + 6$		$y = 2x - 2$
$y = x + 7$	is parallel to	$y = x$
$y = {}^-3x + 7$		$y = {}^-3x$
$y = 5$		$y = {}^-1$

Challenge

These pairs of straight line graphs cut each other at right-angles. Can you spot a rule – check your rule by drawing some more graphs. It's easier if you use a computer or calculator!

This graph	cuts this one at right-angles
$y = {}^-2x + 6$	$y = 0.5x + 5$
$y = {}^-0.25x$	$y = 4x + 1$
$y = 10x$	$y = {}^-0.1x$
$y = x$	$y = {}^-x + 7$

Growth

A1

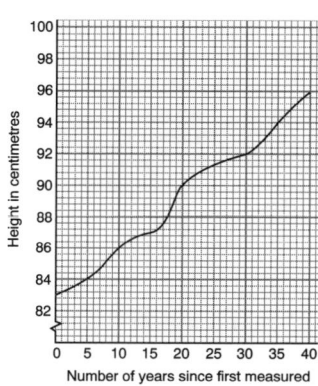

(a) The stalagmite was 88 cm tall 17 years later.

(b) 32 to 33 years after it was first measured it was 93 cm tall.

(c) Over the 40 years it grew 13 cm (96 − 83).

(d) It grew more in the first 20 years compared with the second 20 years.

A2 (a)

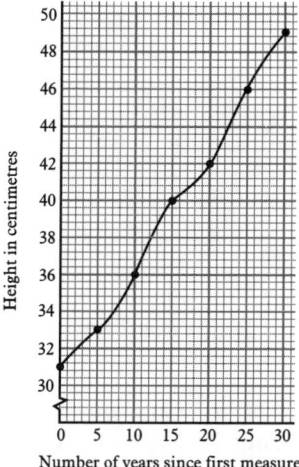

(b) It was 35 cm tall between 8 and 9 years after it was first measured.

A3 (a)

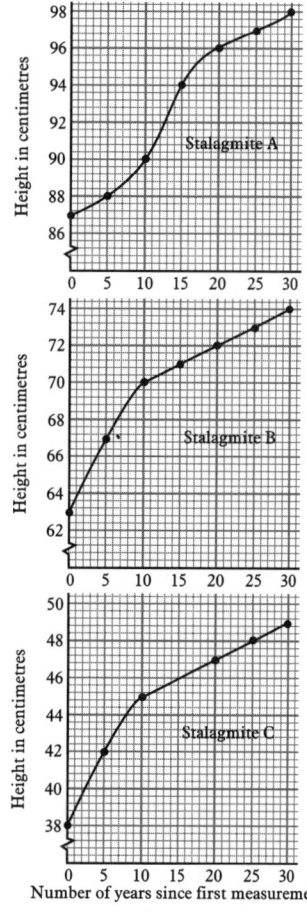

(b) B and C could be in the same cave.

B1 (a) 130 cm

(b) 50 cm (*200 − 70 − 80 or 200 − (70 + 80)*)

B2 (a) 85 cm   (b) 75 cm   (c) 40 cm

B3 24 cm

B4

Number of years since first measured	0	1	2	3	4	5	6
Distance from tip to ground, in cm	130	128	125	124	122	120	118

27

B5 (a)

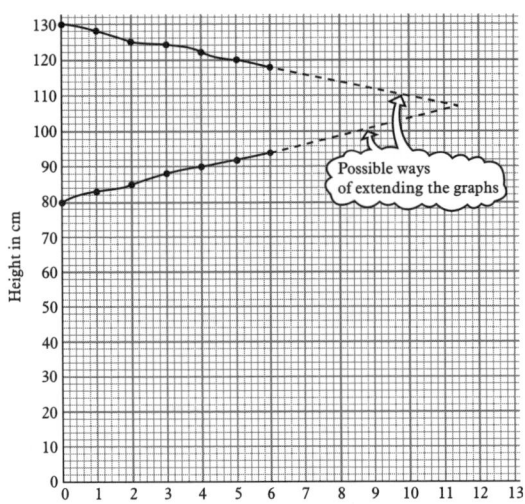

(b) 36 cm (c) About 18 to 22 cm
(d) About 10 to 13 years after first
   measurement

C1

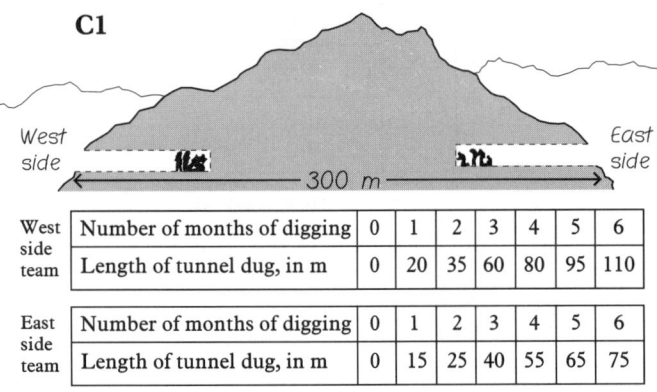

West side team	Number of months of digging	0	1	2	3	4	5	6
	Length of tunnel dug, in m	0	20	35	60	80	95	110

East side team	Number of months of digging	0	1	2	3	4	5	6
	Length of tunnel dug, in m	0	15	25	40	55	65	75

(a) The west side team digs faster.
(b) After six months the west team had
   dug 35 metres more.

C2 (a) After two months digging, 60 metres
     in total had been dug.
   (b) After this time there was still another
     240 metres to be dug.

C3 (a) After 3 months the gap was 200
     metres.
   (b) After 6 months it was 115 metres.

C4 Here is the completed table.

Number of months digging	0	1	2	3	4	5	6
Total length built, in metres	0	35	60	100	135	160	185

C5 (a)

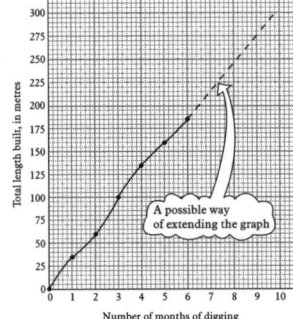

(b) The tunnel should be finished about 9
   to 11 months after the start.

Growth: extension

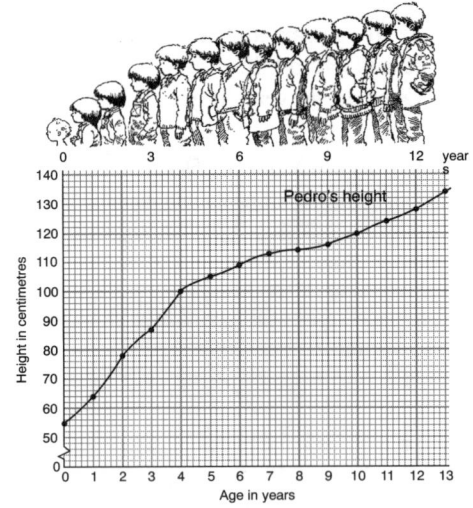

A1 (a) When Pedro was 2 he was 78 cm tall.
   (b) 115 cm at 8·5 years.

A2 (a) He was 90 cm at about 3·3 years.
   (b) At about 11·8 years he was 1·27 m.

A3 When he is grown up he should be about
   1·60 m tall.

A4 (a) At birth he was 55 cm tall.
   (b) He was double his height at birth
     when he was about 6·3 years old.

A5 (a) 9 cm (b) 9 cm (c) 4 cm (d) 4 cm

A6 (a) 45 cm (b) 14 cm
   (c) The graph is steeper between 0 and
     4 years.
   (d) He is growing faster between 0 and
     4 years.
   (e) Between 8 and 9 years he started to
     grow quickly again.

28

B1

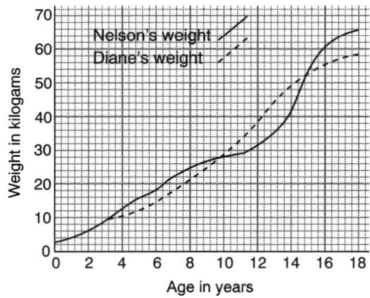

(a) Nelson was heavier at 5.
(b) They were the same weight at 10.
(c) Diane was heavier at 12.
(d) Nelson was heavier at 16.

B2 At 10 and 14·6 years old the twins weighed the same.

B3 (a) The difference in weight was about 3 or 4 kg at 5 years old.
(b) At 13 years their difference in weight was about 8 or 9 kg.

B4 (a) Between age 3 and 10 and after 14·6 Nelson was heavier than Diane.
(b) Between age 10 and 14·6 Diane was heavier than Nelson.

B5 (a) The graph of Nelson's weight becomes steeper after he is 12.
(b) He is not growing so quickly after he is about 16.
(c) About age 10 years Diane started to put weight on fairly quickly.
(d) About age 14 or 15 Diane's increase in weight started to slow down.

B6

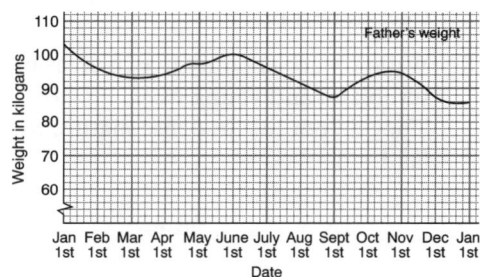

Your description of the graph may include some of these facts.
He weighs 102 kg at the start. During January and February he loses 9 kg. During March, April and May he gains 7 kg but during June, July and August he loses nearly 13 kg. His weight increases by 8 kg in September and October. He loses weight again in November and December and at the end of the year he has lost a total of 16 kg.

B7 Your own sketch graphs. Show them to your teacher.

C1

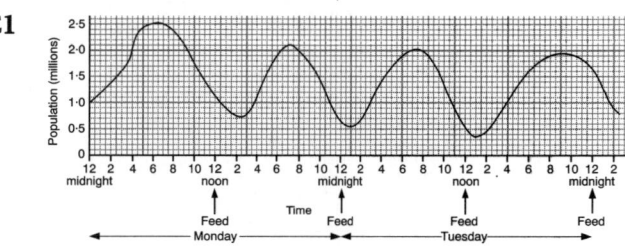

The maximum number of microbes was 2·55 millions.

C2 No, there was a delay of an hour or two after food was given before the microbes started to increase in numbers.

C3 The population did not grow to the same maximum size after each feeding; it was slightly less each time.

C4

Time . . .	The population was . . .
(a) 5 a.m. Monday	Increasing
(b) 9 a.m. Monday	Decreasing
(c) 8 p.m. Monday	Decreasing
(d) 8 a.m. Tuesday	Decreasing
(e) 7 p.m. Tuesday	Increasing
(f) midnight Tuesday	Decreasing

C5 (a) It was growing faster at 4 a.m. Monday.
(b) It was growing faster at 4 a.m. Tuesday.

C6 The delay in the growth after feeding may be due to the time taken to digest the food and for the reproduction process to take place.

D1

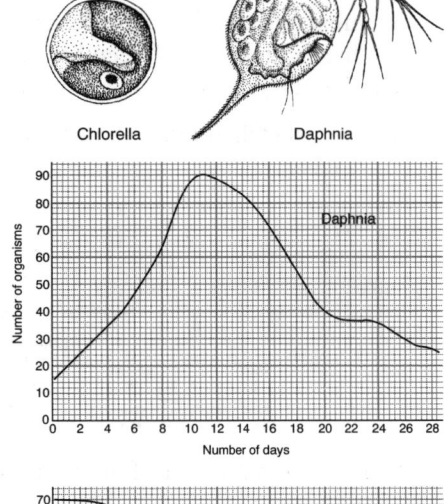

Chlorella   Daphnia

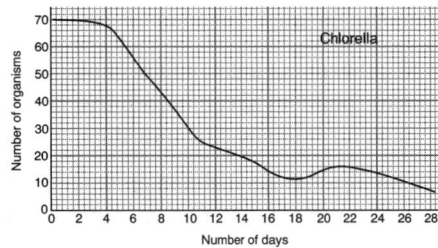

The daphnia increased in number for over 10 days.

D2 The daphnia population decreased fairly slowly for about 3 days, quickly for about 6 days, then slowly.

D3 The chlorella population remained fairly steady for about 4 days, then decreased quickly for 6 days, then slowly for about 8 days. It then increased slightly for 4 days but then decreased again.

D4 After just under 7 days the two populations were the same.

D5 From the graphs you can see that daphnia eat chlorella.
When the daphnia increase in population the chlorella decrease!

D6 To start with the daphnia population rises quickly because it has a large quantity of chlorella organisms to eat. But by the 12th day the daphnia population has eaten two-thirds of the chlorella population. So, if there is nothing else to eat, there is now

not enough food for the daphnia population and it starts to decline.

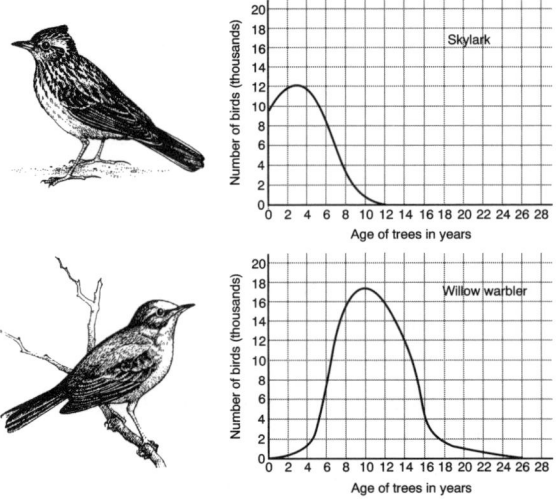

E1 The willow warbler did not live in the area when the trees were first planted.

E2 There were between 9000 and 10000 skylarks when the trees were first planted.

E3 After about 10 years the number of willow warblers was greatest.

E4 The maximum number of willow warblers was about 17000.

E5 Between 6 and 8 years after planting the fall in the number of skylarks was greatest.

E6 After about 6 years the numbers of both species of birds was the same.

E7 The willow warbler is more likely to use trees to nest in.

E8 Willow warblers only live in areas of scrubland and young trees.
They do not live in dense mature woodland.

U-shaped graphs

A1 The area is 250 sq m.

A2

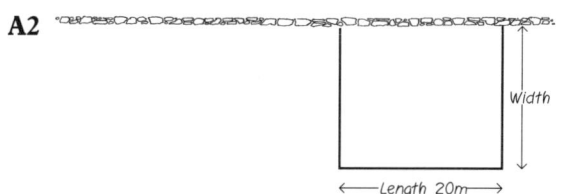

(a) The width of the pen is 20 m
((60 − 20) ÷ 2).

(b) The pen has an area of 400 sq m.

A3 (a) If the pen were 24 m long it would be
18 m wide.

(b) Its area would be 432 sq m.

A4 If the pen were 12 m long it would have an
area of 288 sq m.

A5

Length of pen, in metres	4	8	12	16	20	24	28	32	36	40	44	48
Width of pen, in metres	28	26	24	22	20	18	16	14	12	10	8	6
Area of pen, in sq metres	112	208	288	352	400	432	448	448	432	400	352	288

A6

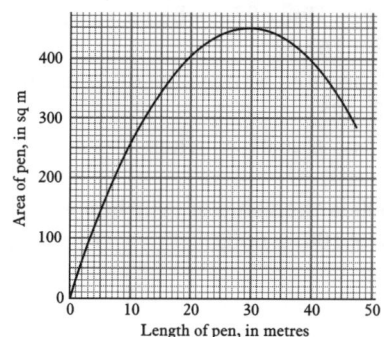

A7 (a) From the graph the largest possible
area is 450 sq m.

(b) For this area the pen should be 30 m
long.

B1

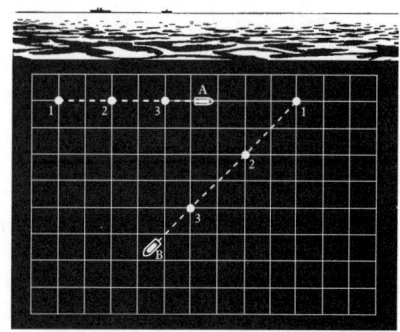

Ship B is travelling faster.

B2 Your own map.

B3 The distance between the ships at
1 o'clock is 90 km.

B4 At 3 o'clock they are 41 km apart.

B5

Time	Distance between ships (km)
1	90
1:30	71
2	54
2:30	42
3	41
3:30	51
4	67
4:30	86
5	106

B6

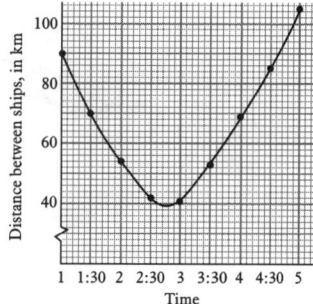

B7 (a) The ships were nearest between 2:40
and 2:50.

(b) They were about 40 km apart at this
time.

B8 The ships were less than 60 km apart
between about 1:50 and 3:50.

C1

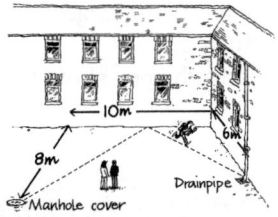

(a) Your own drawing

(b) $11 \cdot 3\,\text{m} + 6 \cdot 3\,\text{m} = 17 \cdot 6\,\text{m}$

C2 The child will run a distance of $17 \cdot 2\,\text{m}$ if he touches a point $4\,\text{m}$ from the corner.

C3

Distance from corner to touching point, in metres	0	2	4	6	8	10
Distance run, in metres	18·8	17·6	17·2	17·4	18·2	19·7

C4 (a) and (b)

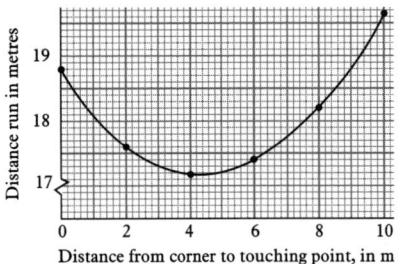

(c) The shortest possible running distance is about $17 \cdot 2\,\text{m}$.

(d) To run this distance they need to touch the wall about $4 \cdot 3$ from the corner.

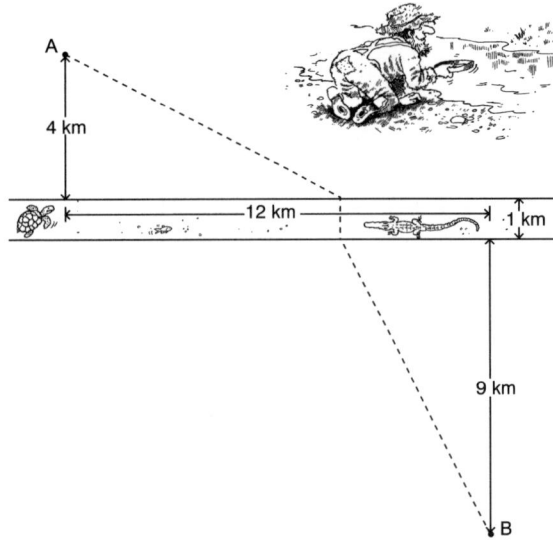

D1 Your own copy of the drawing.

D2

Distance CP in km	Length of journey in km
0	20·0
2	18·9
4	18·7
6	19·0
8	19·8
10	21·0
12	22·6

D3

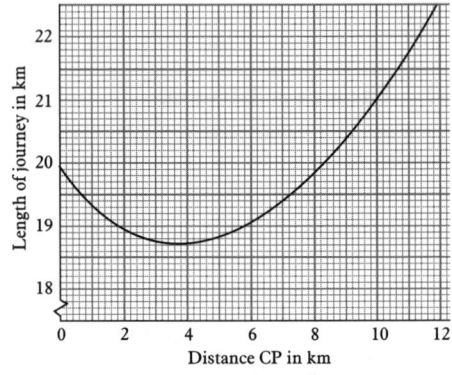

D4 (a) The shortest distance the prospector has to travel is about $18 \cdot 7\,\text{km}$.

(b) He should cross the river about $3 \cdot 7\,\text{km}$ from C.

Formulas and graphs

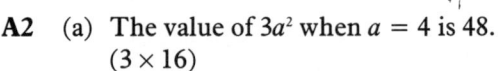

A1 For the formula $d = 5t^2$:

(a) When $t = 2$ then $d = 20$.

(b) When $t = 3$ then $d = 45$.

(c) When $t = 4$ then $d = 80$.

(d) When $t = 10$ then $d = 500$.

A2 (a) The value of $3a^2$ when $a = 4$ is 48.
(3×16)

(b) When $b = 5$ the value of $2b^2$ is 50.
(2×25)

(c) The value of $5x^2$ when $x = 2$ is 20.
(5×4)

(d) 600 is the value of $6y^2$ when $y = 10$.
$(6 \times 100 = 600)$

A3 $s = t^2 + 2t + 6$

When $t = 4$, $s = (4 \times 4) + (2 \times 4) + 6$
$= 16 + 8 + 6 = 30$.
When $t = 5$, $s = (5 \times 5) + (2 \times 5) + 6$
$= 25 + 10 + 6 = 41$.
When $t = 6$, $s = (6 \times 6) + (2 \times 6) + 6$
$= 36 + 12 + 6 = 54$.

A4 $b = 3a^2 + 8$

When $a = 1$, $b = (3 \times 1 \times 1) + 8 = 3 + 8 = 11$.
When $a = 2$, $b = (3 \times 2 \times 2) + 8 = 12 + 8 = 20$.
When $a = 3$, $b = (3 \times 3 \times 3) + 8 = 27 + 8 = 35$.

A5 $f = 4e^2 + 2e$

When $e = 1$, $f = (4 \times 1 \times 1) + (2 \times 1)$
$= 4 + 2 = 6$.
When $e = 2$, $f = (4 \times 2 \times 2) + (2 \times 2)$
$= 16 + 4 = 20$.
When $e = 3$, $f = (4 \times 3 \times 3) + (2 \times 3)$
$= 36 + 6 = 42$.
When $e = 4$, $f = (4 \times 4 \times 4) + (2 \times 4)$
$= 64 + 8 = 72$.

A6 $k = 5h^2 + 3h + 2$
(a) When $h = 2$, $k = 5 \times 4 + 6 + 2 = 28$.
(b) When $h = 3$, $k = 56$.
(c) When $h = 6$, $k = 200$.
(d) When $h = 10$, $k = 532$.

A7 $q = 2p^2 - p$
(a) When p is 5, $q = 45$.
(b) When p is 8, $q = 120$.
(c) When p is 10, $q = 190$.

B1 Here is the table showing h for different values of t. ($h = 25t - 5t^2$)

Time in seconds, t	0	1	2	3	4	5
Height in metres, h	0	20	30	30	20	0

B2 Show your graph to your teacher.

B3 (a) 2·5 seconds from being thrown the stone reaches its greatest height.
(b) Its greatest height reached was about 31 or 32 metres.
(c) The stone first reached a height of 25 m about 1·4 seconds from being thrown.
(d) About 3·6 seconds from being thrown the stone was 25 m high again.
(e) The stone was higher than 15 m for about 3·6 seconds.

C1 Here is the table showing t for various values of s calculated from the formula
$$t = \frac{100}{s}.$$

Speed in m.p.h. (s)	10	20	40	50	60	80	90	100
Time in hours (t)	10	5	2·5	2	1·7	1·3	1·1	1

C2 Your own graph.

C3 (a) 0·1 as a fraction is $\frac{1}{10}$.
(b) 6 minutes ($\frac{1}{10}$ of 60 is 6).
(c) 18 minutes (0·3 is $\frac{3}{10}$).
(d) 2 hours 24 minutes is the same as 2·4 hours (0·4 is $\frac{4}{10}$).

C4 (a) 3·2 hours is the same as 3 hours 12 minutes.
(b) 2 hours 36 minutes is the same as 2·6 hours.
(c) 1 hour 6 minutes is the same as 1·1 hours.
(d) 5 hours 54 minutes is the same as 5·9 hours.
(Remember, $\frac{1}{10}$ of an hour is 6 minutes.)

C5 The answers to these come from your graph.
(a) Between about 6 hours 20 minutes and 7 hours. (Journey at 15 m.p.h.)
(b) About 140 to 180 minutes. (Journey at 25 m.p.h.)
(c) About 1 hour 30 minutes. (Journey at 65 m.p.h.)
(d) About 10 to 15 minutes. (Journey at 75 m.p.h.)

C6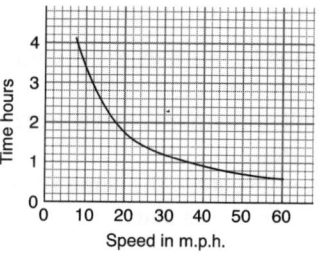

It's probably easiest to find the speed for a travel time of one hour. Why?
The journey was about 35 or 36 miles.

D1 The formula to be used is $h = 4t$, (h is the height in cm, t the time in minutes).
(a) When $t = 0$ then $h = 0$.
(b) When $t = 3$ then $h = 12$.
(c) When t is 5 then $h = 20$.
(d) When t is 8 then $h = 32$.

D2 This question is asking when $h = 40$ what is t? The value of t is 10 minutes.

D3, D4 (b) and **D6**

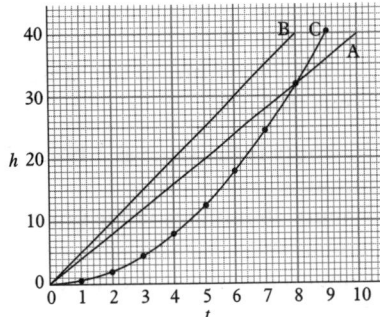

D4 (a) Here are the values of t and h you needed to draw the graph.

Time in minutes, t	0	2	4	6
Height of water in cm, h	0	10	20	30

D5 Here is the completed table for the formula $h = \frac{t^2}{2}$.

Time in minutes, t	0	1	2	3	4	5	6	7	8	9
Height of water in cm, h	0	0·5	2	4·5	8	12·5	18	24·5	32	40·5

D7 About 6·3 minutes from the start. C is more than half full then.

D8 (a) It takes 10 minutes to fill container A.
(b) B takes 8 minutes to fill.
(c) To fill container C takes nearly 9 minutes.

D9 The water levels are the same in A and C at 8 minutes from the start.
(They are also the same – zero – at the start!)

D10 At 5 minutes from the start the height in B is twice the height in C.

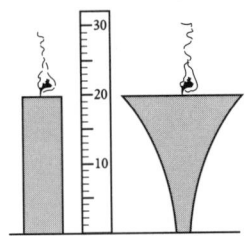

E1 Using the formula $h = 20 - t$:
(a) 15 cm ($t = 5$ hours)
(b) 12 cm ($t = 8$ hours)
(c) 10 cm ($t = 10$ hours)
(d) 5 cm ($t = 15$ hours)

E2 and E5

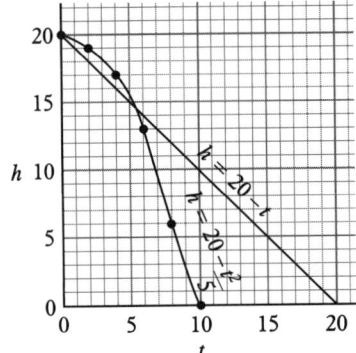

E3 (a) When $t = 3$, $h = 20 - \frac{3 \times 3}{5} = 20 - \frac{9}{5}$
$= 20 - 1·8 = 18·2$
(b) when $t = 4$
$h = 20 - \frac{4 \times 4}{5} = 20 - \frac{16}{5} = 20 - 3·2 = 16·8$

E4 Here is the completed table. The formula was $h = 20 - \frac{t^2}{5}$.

Time in hours, t	0	2	4	6	8	10
Height in cm, h	20	19·2	16·8	12·8	7·2	0

E6 The two candles are the same height when $t = 5$.

E7 (a) When $t = 2$ the straight candle is 18 cm high.
(b) When $t = 7$ it is 13 cm high.
(c) When $t = 16$ it is 4 cm high.

E8 From the graph for the curved candle:
(a) When t is about **3**, h is 18 cm.
(b) When t is about **6**, h is 13 cm.
(c) When t is about **9**, h is 4 cm.

E9 When t is about 8·5 minutes the straight candle is twice the height of the curved one.

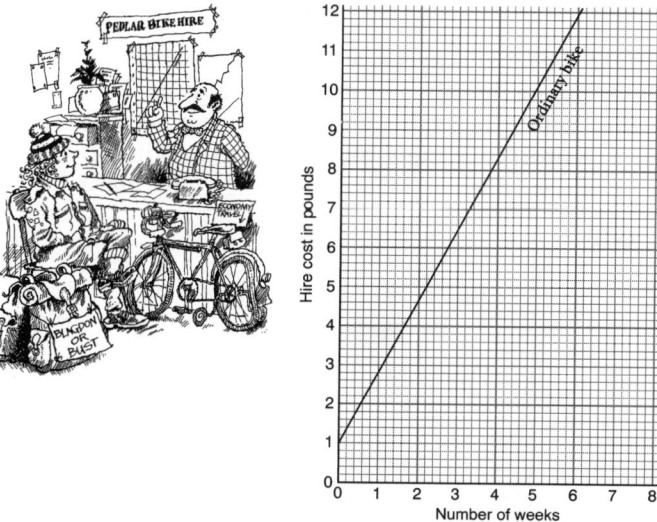

F1 It costs £11 to hire an ordinary bike for 5 weeks.

F2 It costs £2 for each extra week you hire a bike.

F3, F5 and F7

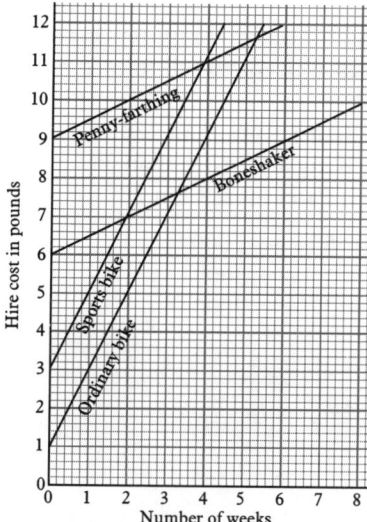

F4 (a) When n is 2, $c = 7$.
    (b) When n is 4, $c = 11$.
    (c) $c = 21$, when n is 9.

F5 The graph for the sports bike is parallel to the one for the ordinary bike.

F6 The formula for hiring an ordinary bike is $c = 2n + 1$, where c is the cost in pounds and n the number of days hired.

F7 Look at the graph.

F8 (a) An ordinary bike is cheapest to hire for 3 weeks.
    (b) A Penny-farthing is the most expensive to hire for 3 weeks.

F9 (a) The cheapest to hire for 5 weeks is a Boneshaker.
    (b) A sports bike is the most expensive to hire for 5 weeks.

F10 When you hire for 2 weeks the Boneshaker and sports bike cost the same.

F11 The Penny-farthing and sports bike cost the same when you hire for 4 weeks.

F12 The graphs cross at $5\frac{1}{3}$ weeks. So if the dealers only hire out bikes for whole numbers of weeks they can never cost the same amount to hire.

F13 Ordinary, Boneshaker, sports, Penny-farthing is the order of cost for a 3-week hire.

F14 Boneshaker, ordinary, Penny-farthing, sports is the order of cost for a 5-week hire.

G1 $P = 100n + 10000$, where P is Angela's pay (£s) and n the number of computers she sells.

G2 (a) If she sells 15 computers she will earn £11500.
    (b) Angela earns £13000 when she sells 30 computers.

G3 (a) 25 computers sold will earn her £12500.
    (b) She will need to sell 37 to earn £13700.
    (c) To earn £30000 Angela will need to sell 200 computers.

G4 Here is the completed table.

Number of computers sold, n	0	30	60	90	120
Pay in pounds, P	10 000	13 000	16 000	19 000	22 000

G5 and **G9**

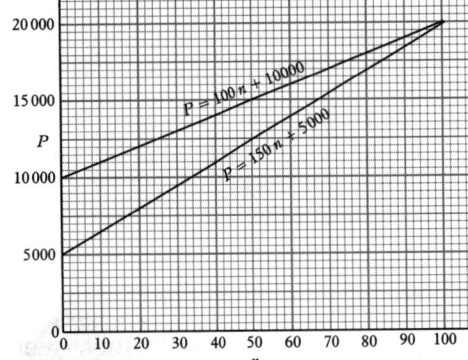

G6 She is paid £5000 per year plus £150 for each computer she sells.

G7 (a) If she sells 15 computers she earns £7250.
(b) If she sells 30 computers Angela earns £9500.

G8 (a) 50 computers   (b) 58 computers
(c) She cannot earn £30 000 exactly, but if she sells 167 computers she will earn £30 050.

G10 Angela needs to sell 100 computers to earn the same amount on the old and new formulas.

G11 Joe Williams is paid £9000 per year plus £50 for each computer he sells.
$$P = 50n + 9000$$
Devon gets £3000 plus £500 per computer.
$$P = 500n + 3000$$

Shailesh earns £6000 plus £200 per computer.
$$P = 200n + 6000$$

G12 The more expensive the computer, the more the salesperson will probably get for selling it. So, Joe Williams probaby sells small computers, Shailesh Patel middle sized ones, and Devon Kingston large ones.

G13

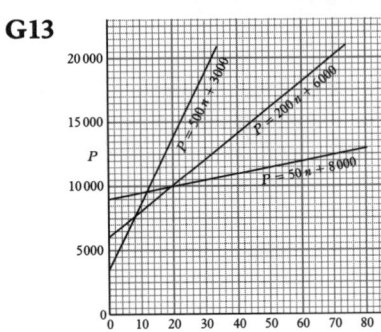

G14 Here is how many computers each salesperson needs to sell to earn more than £15 000.
Joe Williams more than 120
Devon Kingston more than 24
Shailesh Patel more than 45

G15 Here is how many computers each salesperson needs to sell to earn more than £20 000.
Joe Williams more than 220
Devon Kingston more than 34
Shailesh Patel more than 70

G16 (a) $P = 50n + 11\,000$
(b) $P = 550n + 3000$
(c) $P = 225n + 7000$

H1

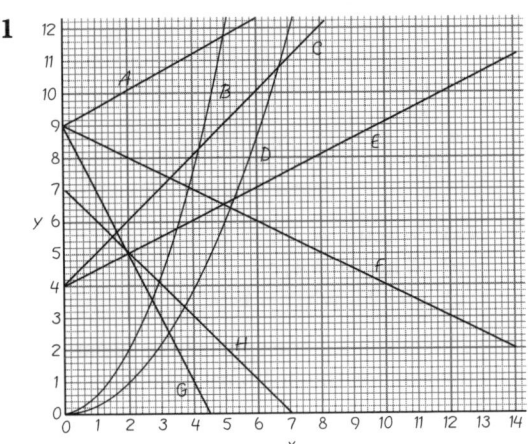

A $y = 9 + \frac{1}{2}x,$ B $y = \frac{1}{2}x^2,$
C $y = 4 + x,$ D $y = \frac{1}{4}x^2,$
E $y = 4 + \frac{1}{2}x,$ F $y = 9 - \frac{1}{2}x,$
G $y = 9 - 2x,$ H $y = 7 - x.$

(How could you convince someone that these answers were correct?)

Travel graphs

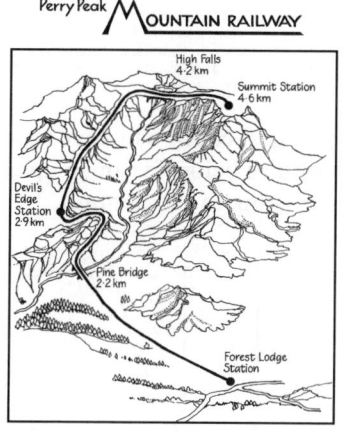

A1 (a) Pine Bridge to Devil's Edge Station is 0·7 km.
(b) Devil's Edge to High Falls is 1·3 km.
(c) High Falls to the Summit Station is 0·4 km.

A2 The average speed of the train from Forest Lodge to Pine Bridge is 19 km/h.
($2·2 \div 7$ km per min; to change into km/h multiply by 60.)

A3 (a) 5 km/h (b) 8 km/h (c) 5 km/h
(Remember, the speeds must be in km per hour.)

B1

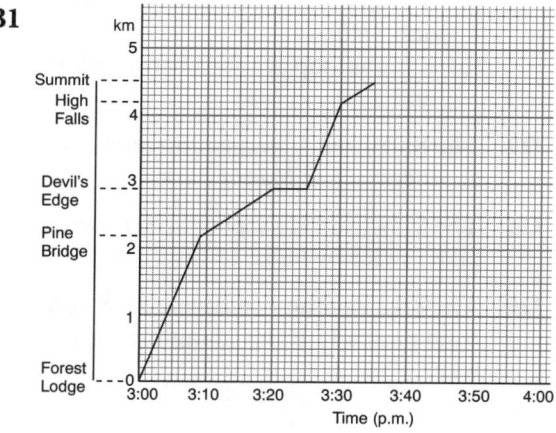

(a) The train passed Pine Bridge at 3:09.
(b) From Pine Bridge to Devil's Edge took 11 minutes.
(c) The train stopped at Devil's Edge between 3:20 and 3:25.

B2 The graph is least steep from Pine Bridge to Devil's Edge, so this means the train is going slowest over this stretch. This suggests that Pine Bridge to Devil's Edge is probably the steepest part of the track.

B3 (a) 15 km/h (b) 4 km/h
(c) 13 km/h (d) 6 km/h

B4, B6 and **B7**

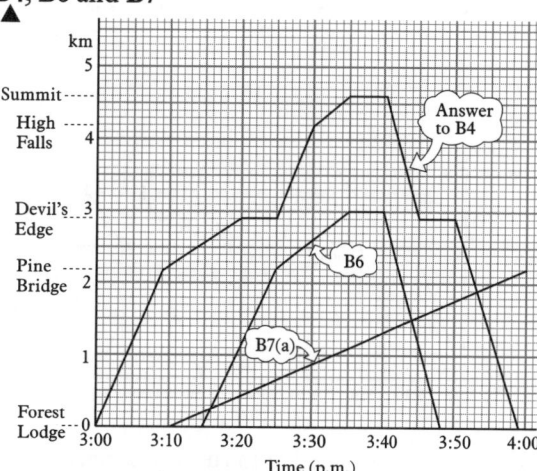

B5 The downward train crosses Pine Bridge at 3:52.

B7 (b) At 3:16, 3:44 and 3:53 (from your graph.)

C1, C2 and **C3**

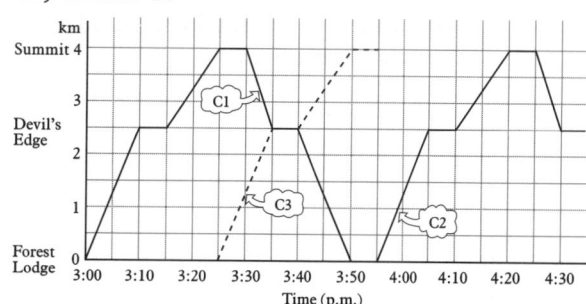

C4 No. If the second train waits 5 minutes at the Summit and then returns it will get to Devil's Edge 5 minutes before the next upward train arrives.

37

C5 You can make upward and downward
△ trains reach Devil's Edge together by
increasing the waiting time at the Summit
to 10 minutes. If you have a different way
of running the service, show it to your
teacher.

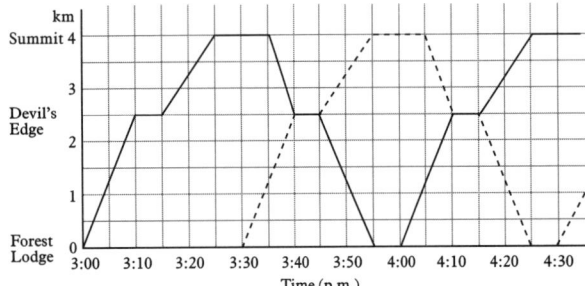

C6
△

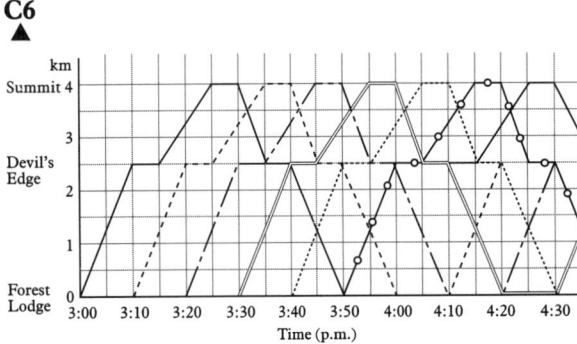

C7 6 trains are needed to run the service.

C8 Trains pass 1·25 km, 2·5 km and 3·5 km
from Forest Lodge.

C9 If you travel to the Summit and back on the
3:30 train from Forest Lodge you will pass
(a) 3 trains on your way up and
(b) 4 trains on the way down (5 if you
count the one leaving Forest Lodge
just as you get back).

C10 Here is the diagram showing the coach
journey.

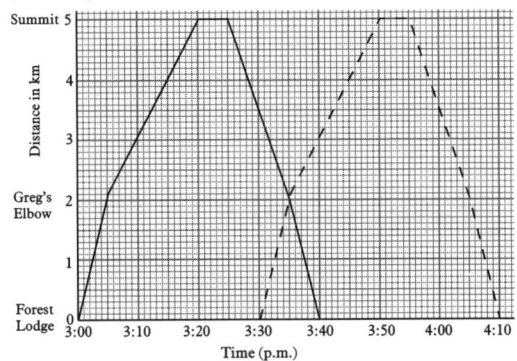

(d) The first coach will have to wait for 20
minutes before it can go up again,
non-stop, to the Summit.

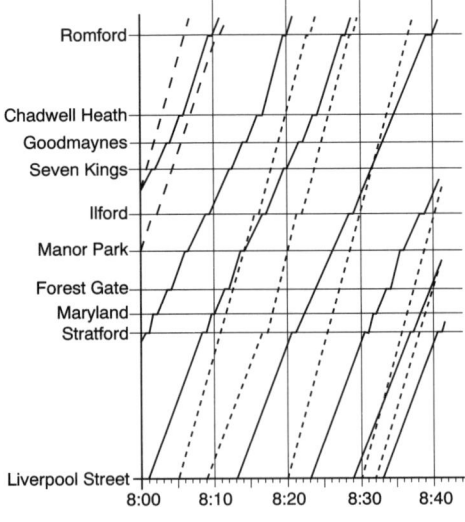

D1 The 8:09 from Liverpool Street stops at
Stratford, Ilford and Romford.

D2 (a) 3 trains stop at Maryland between 8:00
and 8:40.
(b) 7 trains stop at Ilford between 8:00
and 8:40.

D3 The 8:05 from Liverpool Street passes
another train at Manor Park.

D4 The 8:29 from Liverpool Street was over-
taken twice by faster trains.

D5 The 8:12 Romford to Liverpool Street
train passes 11 trains going the opposite
way.

Graphical Representation

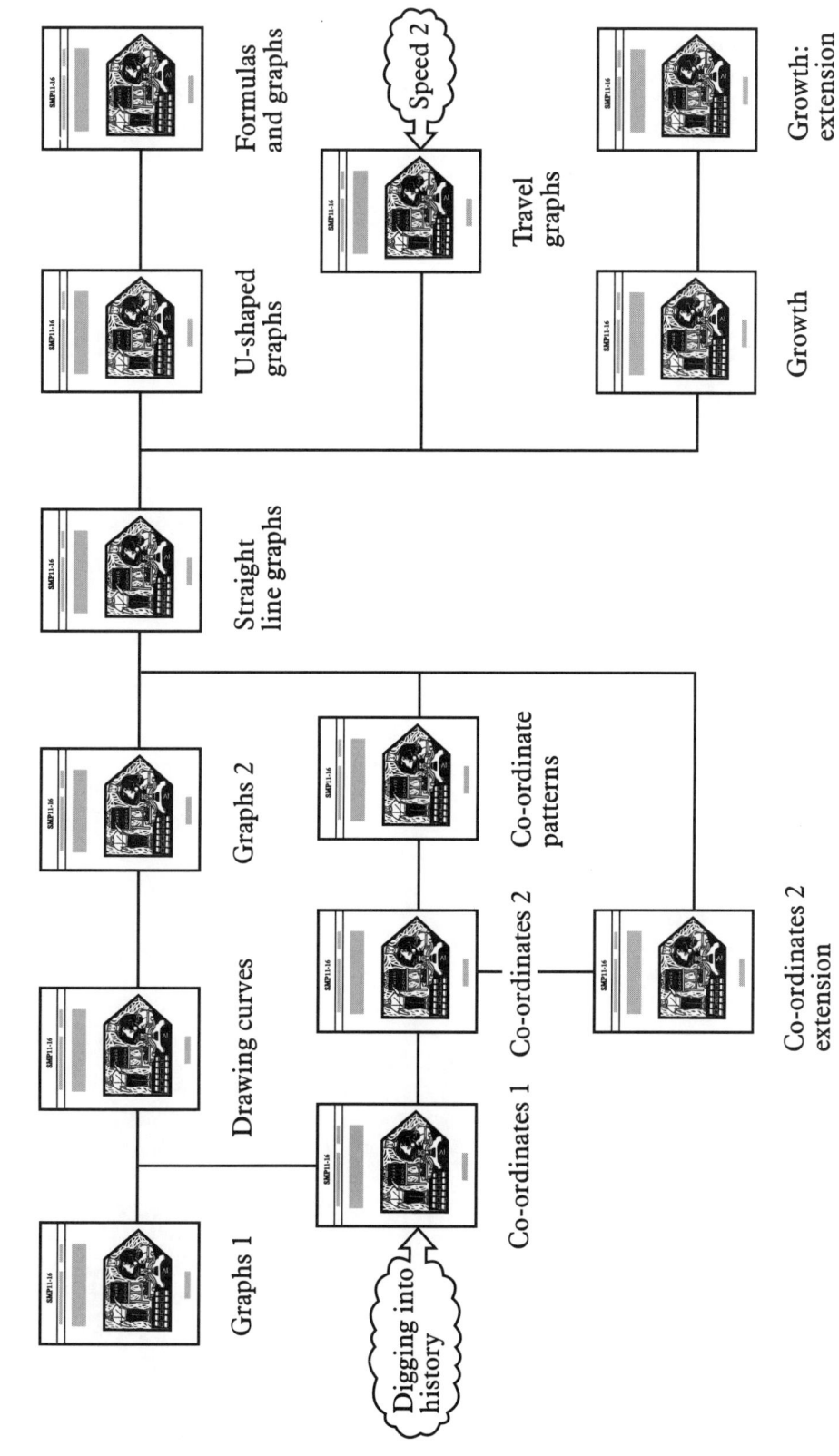

Graphs 1   Drawing curves   Graphs 2   Straight line graphs   U-shaped graphs   Formulas and graphs

Digging into history

Co-ordinates 1   Co-ordinates 2   Co-ordinate patterns

Co-ordinates 2 extension

Travel graphs

Speed 2

Growth   Growth: extension